SERIES EDITOR: BRIAN SEAGER

GRADUATED ASSESSMENT

OCR GCSE MATHEMATICS

STAGES

SECOND EDITION

- Howard Baxter
- Michael Handbury
- John Jeskins
- Jean Matthews
- Mark Patmore

Hodder Murray

A MEMBER OF THE HODDER HEADLINE GROUP

The Publishers would like to thank the following for permission to reproduce copyright material:

Photo credits: p. 17 © Thomas Coex/AFP/Getty Images

Acknowledgements Every effort has been made to trace all copyright holders, but if any have been inadvertently overlooked the Publishers will be pleased to make the necessary arrangements at the first opportunity.

Hodder Headline's policy is to use papers that are natural, renewable and recyclable products and made from wood grown in sustainable forests. The logging and manufacturing processes are expected to conform to the environmental regulations of the country of origin.

Orders: please contact Bookpoint Ltd, 130 Milton Park, Abingdon, Oxon OX14 4SB. Telephone: (44) 01235 827720. Fax: (44) 01235 400454. Lines are open from 9 a.m. to 5 p.m., Monday to Saturday, with a 24-hour message-answering service. Visit our website at www.hoddereducation.co.uk.

© Howard Baxter, Michael Handbury, John Jeskins, Jean Matthews, Mark Patmore, Brian Seager, Eddie Wilde, 2006
First published in 2006 by
Hodder Murray, an imprint of Hodder Education,
a member of the Hodder Headline Group, an Hachette Livre UK company
338 Euston Road,
London, NW1 3BH

Impression number   10 9 8 7 6 5 4 3 2
Year   2011 2010 2009 2008 2007

Cover photo © Andy Sacks/Photographer's Choice/Getty Images
Illustrations © Barking Dog Art
Typeset in Futura Book 12/14pt by Pantek Arts Ltd, Maidstone, Kent
Printed and bound in Singapore

A catalogue record for this title is available from the British Library.

ISBN-13: 978 0340 915 912

Stage 1 Contents

STAGE
1

Contents

STAGE
1

Introduction

About this book

This course has been written especially for students following OCR's Modular Specification C, Graduated Assessment (J516 and J517) for GCSE Mathematics.

This book covers the complete specification for Stages 1 and 2.

- Each chapter is presented in a way which will help you to understand the mathematics, with straightforward explanations and worked examples covering every type of problem.
- At the start of each chapter are two lists, one of what you should already know before you begin and the other of the topics you will be learning about in that chapter.
- 'Activities' offer a more interesting approach to the core content, giving opportunities for you to develop your skills.
- 'Challenges' are rather more searching and are designed to make you think mathematically.
- There are plenty of exercises to work through to practise your skills.
- Some questions are designed to be done without a calculator, so that you can practise for the non-calculator sections of the examination papers.
- Look out for the 'Exam tips' – these give advice on how to improve your performance in the module test, direct from the experienced examiners who have written this book.
- At the end of each chapter there is a short summary of what you have learned.
- Finally, there are 'Revision exercises' at intervals throughout the book to help you revise all the topics covered in the preceding chapters.

Other components in the series

- A Homework Book
  This contains parallel exercises to those in this book to give you more practice.

- An Assessment Pack

  There are two Assessment Packs: one for Foundation Tier (Stages 1 to 7) and one for Higher Tier (Stages 6 to 10). Each contains revision exercises, practice module papers and a practice terminal paper to help you prepare for the examination. Some of the questions in the examination will offer you little help to get started. These are called 'unstructured' or 'multi-step' questions. Instead of the question having several parts, each of which helps you to answer the next, you have to work out the necessary steps to find the answer. There will be examples of this kind of question in the Assessment Pack.

- An Interactive Investigations CD-ROM

  This contains whole-class presentations and individual activities. It helps you understand how you can best use ICT to do your homework and other tasks.

Top ten tips

Here are some general tips from the examiners who wrote this book to help you to do well in your tests and examinations.

Practise

1. **taking time** to work through each question carefully.
2. answering questions **without** a calculator.
3. answering questions which require **explanations**.
4. answering **unstructured** questions.
5. **accurate** drawing and construction.
6. answering questions which **need a calculator**, trying to use it efficiently.
7. **checking answers**, especially for reasonable size and degree of accuracy.
8. making your work **concise** and well laid out.
9. checking that you have **answered the question**.
10. **rounding** numbers, but only at the appropriate stage.

Numbers

1

Whole numbers

There are 26 letters in the English alphabet.

Letters are used to make words.

You can change the order to make a different word.

Changing the order of the letters of the word TRAP gives different words, for example PART and RAPT.

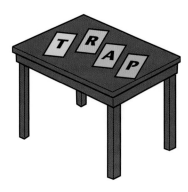

To make numbers there are 10 **digits**.

0 1 2 3 4 5 6 7 8 9

Digits are used to make numbers.

As with letters, you can change the order to make different numbers.

1

Place value

EXAMPLE 1

a) How many different numbers can be made using all of the digits 1, 2 and 3?

Write each number in digits and in words.

b) Which is the largest and which is the smallest of these numbers?

Use the digits systematically.

a)

Hundreds	Tens	Units	In words
1	2	3	one hundred and twenty-three
1	3	2	one hundred and thirty-two
2	1	3	two hundred and thirteen
2	3	1	two hundred and thirty-one
3	1	2	three hundred and twelve
3	2	1	three hundred and twenty-one

b) The largest is 321. The smallest is 123.

EXAM TIP

To write a number in words, always write 'and' after the word 'hundred'.

A ACTIVITY 1

a) Can you find the largest and smallest numbers in Example 1 without writing them all down?

How did you do it?

b) How do you know that you have written down all the possible numbers?

EXAM TIP

To find the largest number, two of the numbers start with 3 in the hundreds position, so the larger digit in the tens position will tell you which is the largest. To find the smallest number, two of the numbers start with 1 in the hundreds position, so the smaller digit in the tens position will tell you which is the smallest.

EXERCISE 1.1

1 How many different numbers can you make using each of these sets of digits?
Write each number in digits.
Write down the largest number and the smallest number made from each set.
a) 2, 8, 7   **b)** 4, 1, 9
c) 5, 5, 8   **d)** 6, 5, 3
e) 8, 9, 7   **f)** 4, 2, 0

2 Write each of these numbers using digits.
a) Three hundred and seventy-five
b) Six hundred and ten
c) Four hundred and seven
d) Eight hundred and fourteen
e) One hundred and thirty
f) One hundred and forty-two
g) Three hundred and seventy-three
h) Two hundred and twelve
i) Six hundred and forty
j) Eight hundred and nine

3 Write each of these numbers in words.
a) 56   **b)** 281
c) 109   **d)** 450
e) 700   **f)** 436
g) 851   **h)** 715
i) 304   **j)** 92

4 Write down the value of each digit shown in **bold**.
a) 4**3**6   **b)** **7**5
c) **2**58   **d)** 5**7**9
e) **4**21   **f)** **5**52
g) 7**9**2   **h)** 30**6**
i) 5**2**4   **j)** **2**90

5 Write each of these sets of numbers in order, smallest first.
a) 483, 834, 348, 384
b) 8795, 5789, 8597, 5879
c) 2643, 2436, 4263, 2364
d) 7218, 8172, 8127, 7821, 7281

Large numbers

Large numbers like 8 765 432 are usually written in groups of three digits, with a small space between each group.

A place value table can help to make sense of the number.

Millions	Hundred thousands	Ten thousands	Thousands	Hundreds	Tens	Units
8	7	6	5	4	3	2

STAGE 1

This tells you that the number is eight million, seven hundred and sixty-five thousand, four hundred and thirty-two.

> **EXAM TIP**
> Notice how the commas in the writing match with the spaces in the number.

A ACTIVITY 2

a) Choose any two digits.

b) Make the largest number and the smallest number using these digits. You must not repeat a digit in either of your numbers.

c) Subtract the smallest number from the largest number.

d) Repeat this for other sets of two digits.

e) What do you notice about your answers?

C CHALLENGE 1

Repeat Activity 2 for numbers made from three digits.

EXERCISE 1.2

1 Write each of these numbers in words.
 a) 2645
 b) 8352
 c) 8526
 d) 5705
 e) 3016
 f) 5540
 g) 2821
 h) 7091
 i) 86151
 j) 80904

2 Write each of these numbers using digits.
 a) Eight thousand, two hundred and fifty-five
 b) Six thousand, eight hundred and fifty-one
 c) Six thousand, nine hundred and twelve
 d) Four thousand, five hundred
 e) Five thousand and fifty
 f) Eight thousand, seven hundred and seventy-two
 g) Five hundred and forty-nine
 h) One thousand, eight hundred and thirteen
 i) Six hundred and forty
 j) Seven hundred thousand, seven hundred

Rounding to the nearest 10, 100, 1000, ...

Often numbers we see are not exact values. They have been **rounded**.

---

25 000 FANS SEE UNITED CUP WIN

It was a capacity crowd that cheered United on Saturday, to a 3–1 win in this year's first round FA Cup competition.

This newspaper headline does not mean that exactly 25 000 fans saw the match. It was probably slightly more or slightly less than this.

Giving an exact value, such as 24 891, would not make such a dramatic headline.

EXAMPLE 2

Round 2376 to the nearest

a) 10. **b)** 100. **c)** 1000.

a) 2376 is between 2370 and 2380.

2376 is nearer to 2380.

b) 2376 is between 2300 and 2400.

2376 is nearer to 2400.

c) 2376 is between 2000 and 3000.

2376 is nearer to 2000.

A number exactly halfway is always **rounded up**.
So 2500 to the nearest 1000 would be 3000 and 2350 to the nearest 100 would be 2400.

EXERCISE 1.3

1 Rewrite each of these newspaper headlines using a suitable approximate number.
Say what accuracy you have used in each case.
  a) 41 638 see Rovers win!
  b) 4327 at pop concert!
  c) MP elected with a 14 873 majority!
  d) Exports exceed £4 683 421!
  e) Bank makes £12 321 457 profit!
  f) Local man wins £998 321 on the lottery!

2 Round each of these numbers
  (i) to the nearest 10.
  (ii) to the nearest 100.
  a) 673   **b)** 451
  c) 333   **d)** 1577
  e) 469   **f)** 4724
  g) 9689   **h)** 5435

3 Round each of these numbers
  (i) to the nearest 10.
  (ii) to the nearest 100.
  (iii) to the nearest 1000.
  a) 1066   **b)** 23 629
  c) 8912   **d)** 26 788
  e) 46 950

4 Round each of these numbers to the nearest million.
  a) 6 299 888
  b) 3 500 000

5 The number 5637 can be written as
  a) 6000
  b) 5600
  c) 5640
  What is the accuracy used in each of these rounded values?

Odd and even numbers

The counting numbers are made up of two sets of numbers.

1	3	5	7	9	11	13

2	4	6	8	10	12

Each of these sets goes up in 2s.

> The numbers 1, 3, 5, 7, 9, 11, 13, ... are called **odd numbers** and
> the numbers 2, 4, 6, 8, 10, 12, ... are called **even numbers**.

Notice how all the even numbers end in 2, 4, 6, 8 or 0 and all the odd numbers end in 1, 3, 5, 7 or 9.

So   429 is an odd number since it ends in 9
and   150 is an even number since it ends in 0.

EXERCISE 1.4

1 List all the odd numbers between
a) 20 and 40.
b) 55 and 70.

2 List all the even numbers between
a) 36 and 53.
b) 26 and 41.

3 Make lists of the odd and the even numbers that are in
a) the 3 times table.
b) the 4 times table.
c) the 5 times table.
d) the 6 times table.

4 Is there a times table where all the answers are odd?

5 Is there a times table where all the answers are even?

A ACTIVITY 3

a) Add two even numbers together.
Do this several times with different even numbers.
What do you notice about your answers?

b) Add two odd numbers together.
Do this several times with different odd numbers.
What do you notice about your answers?

c) Add one odd and one even number together.
Do this several times with different odd and even numbers.
What do you notice about your answers?

5s and 10s

The 5 times table goes   5, 10, 15, 20, 25, 30, 35, 40, ...   and

the 10 times table goes   10, 20, 30, 40, 50, 60, 70, 80, ...

Notice how the numbers in the 5 times table all end in 0 or 5 and the numbers in the 10 times table all end in 0.

We say that the numbers in the 5 times table are **divisible** by 5 and the numbers in the 10 times table are divisible by 10.

||| EXERCISE 1.5

1 Look at this list of numbers.

55 67 150 38 400 125

Write down
a) the numbers which are divisible by 5.
b) the numbers which are divisible by 10.

2 Look at this list of numbers.

60 245 2000 130 50 105

Write down
a) the numbers which are divisible by 5.
b) the numbers which are divisible by 10.

3 Look at this list of numbers.

52 305 210 47 16 100 78 35

Write down
a) the numbers which are odd.
b) the numbers which are even.
c) the numbers which are divisible by 5.
d) the numbers which are divisible by 10.

4 Look at this list of numbers.

625 14 30 58 115 710 1000 503
Write down
a) the numbers which are odd.
b) the numbers which are even.
c) the numbers which are divisible by 5.
d) the numbers which are divisible by 10.

Addition and subtraction

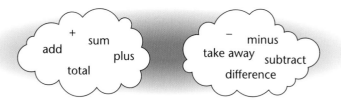

There are various 'mental' ways to do simple addition and subtraction calculations. You may have seen these before.

EXAMPLE 3

Work out these.

a) 23 + 36

b) 58 + 34

a) Since 36 is 30 + 6, add the two parts on separately.
23 + 30 = 53   53 + 6 = 59

b) Similarly,
58 + 30 = 88   88 + 4 = 92

EXERCISE 1.6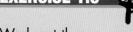

1 Work out these.
 a) 26 + 47   **b)** 24 + 31
 c) 38 + 53   **d)** 52 + 36
 e) 49 + 24   **f)** 26 + 69
 g) 63 + 29   **h)** 43 + 28
 i) 46 + 36   **j)** 86 + 26

2 Add 16 and 22.

3 Add 61 and 25.

4 What is 39 plus 24?

5 What is 74 plus 68?

6 What is the sum of 69 and 31?

7 What is the sum of 54 and 17?

8 Find the total of 57 and 42.

9 Find the total of 39 and 24.

10 Add together 24 and 55.

11 Add together 74 and 22.

EXAMPLE 4

Work out 55 – 28.

Since 28 is 20 + 8, first take away 20, then take away 8.
55 – 20 = 35   35 – 8 = 27

EXERCISE 1.7

1 Work out these.
  a) 39 – 23   **b)** 57 – 14
  c) 86 – 21   **d)** 38 – 25
  e) 48 – 27   **f)** 74 – 53
  g) 73 – 54   **h)** 83 – 39
  i) 55 – 39   **j)** 72 – 19

2 What is 59 take away 26?

3 What is 46 take away 28?

4 Find 86 minus 69.

5 Find 68 minus 43.

6 What is the difference between 80 and 13?

7 What is the difference between 94 and 73?

8 What is 93 subtract 45?

9 What is 83 subtract 78?

10 Work out 72 take away 48.

11 Work out 70 take away 37.

In Example 4,   55 – 28 = 27.

Notice also that   55 – 27 = 28

and that   27 + 28 = 55.

The three numbers are linked together and can make three different calculations.

Knowing the answer to one calculation helps you work out the answers to two others.

EXERCISE 1.8

Use each of these sets of numbers to make three different calculations using + or −.

1 46   28   18   **6** 17   84   67

2 41   56   15   **7** 18   37   19

3 79   91   12   **8** 43   62   19

4 56   19   75   **9** 100   28   72

5 12   24   36   **10** 100   19   81

EXAMPLE 5

How many do you need to add to

a) 61 to make 100?

b) 24 to make 55?

a) First add to get to the next ten.   61 + 9 = 70
  Then add tens to get to 100.   70 + 30 = 100

  9 + 30 = 39

  Or find 100 − 61.

  100 − 60 = 40   40 − 1 = 39

b) First add to get to the next ten.   24 + 6 = 30
  Then add tens to get to 50.   30 + 20 = 50
  Then add units to get to 55.   50 + 5 = 55

  6 + 20 + 5 = 31

  Or find 55 − 24.

  55 − 20 = 35   35 − 4 = 31

EXERCISE 1.9

1 How many do you need to add to
   a) 34 to make 100?   **b)** 16 to make 100?
   c) 58 to make 100?   **d)** 77 to make 100?
   e) 17 to make 100?   **f)** 18 to make 100?
   g) 23 to make 65?   **h)** 24 to make 73?
   i) 49 to make 98?   **j)** 51 to make 93?

Multiplication and division

Finding the answer to any multiplication or division calculation depends upon knowing your times tables.

If you still do not know all your tables by heart, now is the time to learn them.

You can use this multiplication grid to help. Cover it up and try to answer the questions without it.

×	1	2	3	4	5	6	7	8	9	10
1	1	2	3	4	5	6	7	8	9	10
2	2	4	6	8	10	12	14	16	18	20
3	3	6	9	12	15	18	21	24	27	30
4	4	8	12	16	20	24	28	32	36	40
5	5	10	15	20	25	30	35	40	45	50
6	6	12	18	24	30	36	42	48	54	60
7	7	14	21	28	35	42	49	56	63	70
8	8	16	24	32	40	48	56	64	72	80
9	9	18	27	36	45	54	63	72	81	90
10	10	20	30	40	50	60	70	80	90	100

Look at the line from the top left-hand corner to the bottom right-hand corner. The numbers are the same on both sides of the line. Can you see why?

The order in which the numbers are multiplied does not change the answer.

8 × 7 = 56   and   7 × 8 = 56.

EXERCISE 1.10

Copy these calculations and write down the answers.

1	3 × 4	**16**	8 × 5
2	2 × 3	**17**	7 × 4
3	6 × 2	**18**	2 × 2
4	7 × 6	**19**	8 × 5
5	9 × 5	**20**	3 × 7
6	4 × 8	**21**	6 × 2
7	8 × 6	**22**	6 × 4
8	9 × 9	**23**	9 × 4
9	4 × 7	**24**	8 × 8
10	5 × 4	**25**	10 × 5
11	3 × 9	**26**	10 × 3
12	4 × 3	**27**	8 × 6
13	6 × 6	**28**	5 × 7
14	9 × 6	**29**	7 × 7
15	5 × 3	**30**	8 × 9

STAGE
1

Unlike multiplication, there are different ways of writing divisions.

$42 \div 7$, $\frac{42}{7}$ and $7\overline{)42}$ are three ways to write 'divide 42 by 7'.

A multiplication grid can be helpful, but knowing your tables by heart is much better.

×	1	2	3	4	5	6	7	8	9	10
1	1	2	3	4	5	6	7	8	9	10
2	2	4	6	8	10	12	14	16	18	20
3	3	6	9	12	15	18	21	24	27	30
4	4	8	12	16	20	24	28	32	36	40
5	5	10	15	20	25	30	35	40	45	50
6	6	12	18	24	30	36	42	48	54	60
7	7	14	21	28	35	42	49	56	63	70
8	8	16	24	32	40	48	56	64	72	80
9	9	18	27	36	45	54	63	72	81	90
10	10	20	30	40	50	60	70	80	90	100

For $42 \div 7$ you look for 42 in the '7' row. The number at the top of the column, above the 42, shows the answer.

So $42 \div 7 = 6$.

A ACTIVITY 4

Copy and complete this multiplication grid.

×	2	5	3	8	6
4					
2					
5					
9					
6					

A   ACTIVITY 5

a) Multiply two even numbers together.
Do this several times with different even numbers.
What do you notice about your answers?

b) Multiply two odd numbers together.
Do this several times with different odd numbers.
What do you notice about your answers?

c) Multiply one odd and one even number together.
Do this several times with different odd and even numbers.
What do you notice about your answers?

||| EXERCISE 1.11

Work out these.

1 $48 \div 6$

2 $14 \div 2$

3 $4\overline{)28}$

4 $9\overline{)36}$

5 $\dfrac{12}{3}$

6 $\dfrac{21}{7}$

7 $36 \div 9$

8 $70 \div 10$

9 $54 \div 6$

10 $\dfrac{24}{4}$

11 $\dfrac{16}{2}$

12 $27 \div 9$

13 $7\overline{)63}$

14 $40 \div 5$

15 $35 \div 7$

16 $\dfrac{18}{2}$

17 $\dfrac{72}{8}$

18 $8\overline{)64}$

19 $48 \div 8$

20 $\dfrac{45}{5}$

STAGE
1

Just as with addition and subtraction, multiplication and division calculations are linked together.

$$42 \div 7 = 6 \qquad 42 \div 6 = 7 \qquad 6 \times 7 = 42$$

Knowing the answer to one calculation helps you work out the answers to the others.

EXERCISE 1.12

Use each of these sets of numbers to make three different calculations using $\times$ or $\div$.

1	5	7	35		**5**	24	6	4		**8**	8	72	9
2	7	4	28		**6**	8	2	4		**9**	9	36	4
3	63	9	7		**7**	6	48	8		**10**	6	18	3
4	9	45	5									

K KEY IDEAS

- To round a number to the nearest 10, find the number above and the number below it when counting in tens and decide which it is nearer.

- When a number is halfway between the numbers above and below it, it is rounded up.

- Odd numbers end in 1, 3, 5, 7 or 9.

- Even numbers end in 2, 4, 6, 8 or 0.

- Numbers which are divisible by 5 end in 0 or 5.

- Numbers which are divisible by 10 end in 0.

- You should know the times tables up to 10×10.

- Addition and subtraction calculations are linked.

- Multiplication and division calculations are linked.

Probability 2

You will learn about

- The words used to describe the probability of something happening

You should already know

- How to use a simple scale

What is the chance of it raining sometime in the next three weeks?

If that question is asked in the UK, then the answer is probably 'very likely', even if it is the middle of summer! In the UK, even in summer, there is often some rain over a period as long as three weeks.

In many other countries, the answer will be very different.
In extremely dry areas of the world, the answer may be 'very unlikely'. In others, the answer may be 'about evens'. By 'evens' we mean there is the same chance of something happening as not happening. In some countries, the answer may well depend on what season it is when the question is asked, as long periods of wet weather are often followed by long periods of dry weather.

However, wherever the question is asked, the answer will be on a scale like the one below.

impossible very unlikely unlikely evens likely very likely certain

The words on the scale can be used to descibe the chance, or **probability**, of an event happening.

The word 'probability' is used instead of the word 'chance' in mathematics.

EXAMPLE 1

Use probability words such as 'unlikely' and 'certain' to describe the probability of each of these events.

a) A fair coin coming down 'heads' when it is tossed.

b) A cow jumping over the moon.

c) Getting mathematics homework this week.

d) Getting an odd number when a fair dice is thrown.

e) It will get dark tonight.

f) United will score five goals in their next match.

a) Evens: if a coin is fair, the chances of it showing heads and tails are equal.

b) Impossible: this will never happen.

c) This will depend on your school. The answer is probably certain or very likely.

d) Evens: if a dice is fair, the probabilities of it showing each of the numbers are equal and half of the numbers on a dice are odd numbers.

e) Certain: in the UK it gets dark every night.

f) Very unlikely: five goals is a lot of goals so they probably won't score that many, but it is not impossible.

ACTIVITY 1

The probability washing line

- Put a piece of string, about 3 metres long, across the front of the classroom, either from wall to wall or across the face of the front wall. Write the words 'impossible', 'very unlikely', 'unlikely', 'evens', 'likely', 'very likely' and 'certain' on pieces of paper and attach them at equal intervals along the string.

- Each member of the class writes on a piece of paper a description of an event that has not yet happened, for example 'Tomorrow it will rain'.

- Decide as a class where each piece of paper should go.

EXERCISE 2.1

1 Choose an expression from the box to describe the probability of each of the events below.

impossible	very unlikely	unlikely	evens	likely	very likely	certain

   a) Christmas Day will be on 25 December this year.
   b) It will rain every day in June.
   c) The next baby born at your local hospital will be a girl.
   d) The next lorry you see will have a male driver.
   e) The traffic lights are showing blue.

2 Use probability words such as 'unlikely' and 'certain' to describe the probability of each of these events.
   a) The next vehicle to go past your school will be a car.
   b) A ball thrown in the air will come down.
   c) You will get a car for your seventeenth birthday.
   d) You will go to the moon this year.
   e) The sun will be shining when you wake up tomorrow.

STAGE
1

3 Use probability words to describe the probability of each of these events.
   a) Getting tails when you throw a fair coin.
   b) A football team from the top division will win the cup.
   c) You will get a 7 when you throw an ordinary fair dice.
   d) It will snow in England on 1 June.
   e) A person selected at random from your school is left handed.

4 Use probability words to describe the probability of each of these events.
   a) You will get maths homework tonight.
   b) The winning numbers in the next National Lottery draw will be 1, 15, 23, 29, 35, 46.
   c) This lesson will finish before 6 p.m.
   d) The next car that passes the school will have four passengers.
   e) When you take a card from a pack of ordinary playing cards you will get a red card.

5 Use probability words to describe the probability of each of these events when you roll a fair six-sided dice.
   a) You will get a 0.
   b) You will get a 3, 2 or 1.
   c) You will get a 6 or less.
   d) You will get an even number.
   e) You will get a 4 or a 5.

6 Use probability words to describe the probability of each of these events when you take a card from an ordinary pack of playing cards.
   a) You will get a red card.
   b) You will get a blue card.
   c) You will get an ace.
   d) You will get a heart, spade, diamond or club.
   e) You will get a number card.

7 a) A box contains 12 counters.
      Each counter is green, blue or yellow.
      A counter is taken from the box at random.
      If there is an evens chance of getting a green counter, how many green counters are there?
   b) Another box contains 12 counters.
      There are 5 red counters and 7 black counters.
      One counter is taken from the box without looking.
      (i) Which colour is it more likely to be? Explain why.
      (ii) How many of each colour should be removed from the box to make it an evens chance that either colour is chosen?

EXERCISE 2.1 continued

8 I have the eight cards shown.

Two cards are already turned over.
I turn them all over, shuffle the cards and then pick one.
What numbers would the unknown cards be if
a) the probability of picking a 1 is evens?
b) the probabilities of picking a 1 or a 2 are the same?

9 Write down two events which you think are very likely to happen tomorrow, and two which are unlikely to happen.

10 Here is a fair spinner.

a) Copy the diagram and colour the spinner so that there is an evens chance of it landing on black or white.
b) How many different ways can the spinner be coloured so that the probability of landing on black or white is evens?

11 Here is another fair spinner.

For each part of the question, copy the diagram and colour the spinner for the probability to be correct.
a) It is certain to land on black.
b) It is unlikely to land on black.
c) It is impossible to land on black.
d) It is very likely to land on black.

STAGE
1

12 Each of the jars below contains black and white beads only.
For each part of the question, copy the diagram and colour the beads as
necessary for the probability to be correct.
a) The probability of choosing a black is 'certain'.

b) There is an evens chance of choosing a black.

c) The probability of choosing a black is 'unlikely'.

d) The probability of choosing a black is 'very likely'.

K **KEY IDEAS**

■ You can use words such as 'impossible', 'unlikely', 'evens', 'likely' and
'certain' to describe the probability of an event happening.

STAGE
1

Revision exercise A1

1 a) Make as many different numbers as you can using all of the digits 4, 9 and 7.
  b) Write down the largest number and the smallest number made.

2 a) Write the number five thousand and thirty-seven using digits.
  b) Write the number 4719 in words.

3 Round 7835
  a) to the nearest 10.
  b) to the nearest 100.
  c) to the nearest 1000.

4 List all the odd numbers between 16 and 28.

5 List all the even numbers between 121 and 137.

6 List all the numbers between 61 and 93 that are divisible by 5.

7 List all the numbers between 207 and 315 that are divisible by 10.

8 Work out these.
  a) 56 + 8   **b)** 23 + 44
  c) 37 + 15   **d)** 64 + 36
  e) 52 + 28   **f)** 67 − 5
  g) 95 − 14   **h)** 62 − 23
  i) 84 − 27   **j)** 73 − 65

9 Work out these.
  a) 2×3 **b)** 7×5
  c) 6×6 **d)** 4×9
  e) 8×7 **f)** $30 \div 6$
  g) $9\overline{)27}$ **h)** $\frac{80}{10}$
  i) $45 \div 5$ **j)** $\frac{32}{4}$

10 Choose an expression from the box to describe the probability of each of the events below.

impossible	very unlikely	unlikely
evens	likely	very likely	certain

  a) You will go to school on Christmas Day.
  b) A horse will win the next Grand National.
  c) Somebody in your class will get more than 50% in the next mathematics test.
  d) A fair coin will come down heads when it is tossed.
  e) It will rain sometime in the next two weeks.
  f) The school team will win their next hockey match 7–0.

11 You have a fair eight-sided spinner numbered 1 to 8.
Use probability words to describe the probability of each of these things happening when it lands.
a) It lands on an odd number.
b) It lands on a number less than 7.
c) It lands on the number 8.
d) It lands on the number 10.

12 The six-sided spinner shown in the diagram has red, blue and yellow sections.

a) Write down, from most likely to least likely, the order of the sections on which the spinner will land. Give a reason for your answer.
b) Which section of the spinner should be changed to make the chances of its landing on any of the colours equal?

TAGE

1

Direction and position

3

STAGE
1

You will learn about

- Compass directions
- How to use coordinates

You should already know

- Which direction is clockwise and which is anticlockwise
- How to measure distances in centimetres
- How to use a number line

Compass directions

The diagram shows the main compass directions.

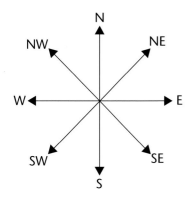

EXAM TIP

To help you remember the four main compass directions you can use the saying '**N**ever **E**at **S**hredded **W**heat'. The first letter of each word gives you the direction, starting at north and going clockwise.

N is north, **E** is east, **W** is west and **S** is south.

EXAMPLE 1

This is a sketch map of an island. North is marked.

a) Which letter is north of C?

b) Which letter is west of C?

c) What is the direction from A to B?

d) What is the direction from F to D?

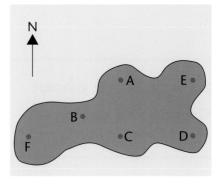

a) A

b) F

c) South west (SW)

d) East

> **EXAM TIP**
> North is usually drawn pointing up the page.

EXERCISE 3.1

Use centimetre-square paper for questions **1** to **4**. Draw the north lines pointing up the page.

1 Mark two points, A and B, where AB is 4 cm long and A is west of B.

2 Mark two points, B and C, where BC is 3 cm long and C is south west of B.

3 a) Mark three points, A, B and C, where A is 4 cm north of B and C is 4 cm west of B.
   b) What is the direction from A to C?

4 a) Mark three points, A, B and C, where A is 2 cm west of B and C is 2 cm south of A.
   b) What is the direction from C to B?

5 Look at this grid.

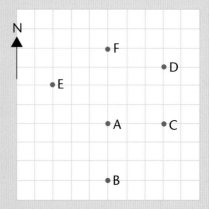

Name the letter that is
a) south of A.
b) north east of A.
c) south west of C.

6 Look at this grid.

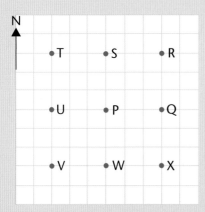

Name the letter that is
a) south of P.
b) north west of W.
c) south east of P.

7 Look at this grid.

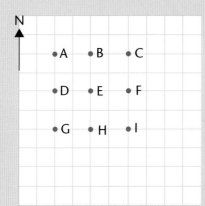

Write down the direction
a) from A to B.
b) from C to G.
c) from E to B.
d) from I to E.

8 Look at this map.

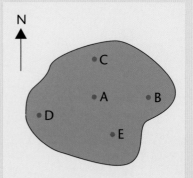

Write down the direction
a) from A to B.
b) from C to D.
c) from E to B.

9 Look at this map.

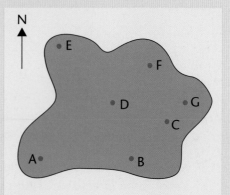

a) Write down the direction
   (i) from A to B.
   (ii) from B to C.
b) Which letter is
   (i) south east of F?
   (ii) north east of C?

EXERCISE 3.1 continued

10 This is a map of part of Carlisle.

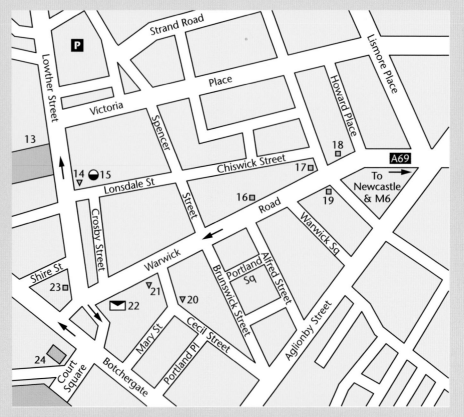

a) Name a road that is roughly north-west to south-east in direction.
b) Debbie cycled from building 14 towards building 17.
  Roughly what direction was that?

11 Monyash is south east of Flagg.
  What is the direction of Flagg from Monyash?

STAGE
1

 ACTIVITY 1

Look at the diagram in Exercise 3.1 question **7** again.

a) How many ways are there to get from A to E?
  List them, for example 'Go east to B, then go south to E'.

b) How many ways are there to get from G to F?

ACTIVITY 2

Mark the four main compass points on the walls of the classroom.

Take turns to direct a student around the room using compass directions only.

ACTIVITY 3

Get an Ordnance Survey map of your local area.

Find and write down the compass directions between different places on the map.

Coordinates

On this grid two lines are drawn.
These are called the **axes**.

The horizontal one is called the **x-axis** and the vertical one is called the **y-axis**.

The lines of the grid are numbered from 0 to 8 across and from 0 to 8 up.

Rather than putting two zeros where the two axes cross, one zero is placed there.

This point is called the **origin**.

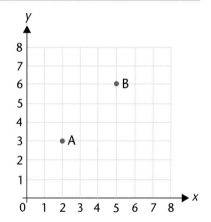

To find a point on a grid, the **horizontal** position (across) is stated first and then the **vertical** (upward) position.

The point marked A is on the line 2 across and 3 up. This is written as (2, 3). These are the **coordinates of A**.

The point marked B is 5 across and 6 up, so its coordinates are (5, 6).

The origin is 0 across and 0 up, up so its coordinates are (0, 0).

EXAMPLE 2

State the coordinates of the points marked P, Q and R.

P is 1 across and 3 up so its coordinates are (1, 3).

Q is 5 across and 1 up so its coordinates are (5, 1).

R is 3 across and 0 up so it is on the x-axis. Its coordinates are (3, 0).

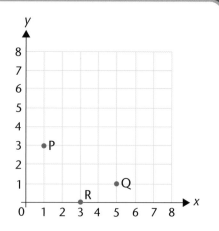

EXAM TIP

A common error when reading graphs is to confuse, for example, (0, 2) and (2, 0).

EXAMPLE 3

On the grid, mark and label the points A(8, 4), B(4, 7) and C(0, 6).

Each point can be plotted with a dot or a cross, and the letter is written next to it.

To plot point A you move 8 along the x-axis and then 4 up.

To plot point B you move 4 along the x-axis and then 7 up.

To plot point C you move 0 along the x-axis and then 6 up. It is on the y-axis.

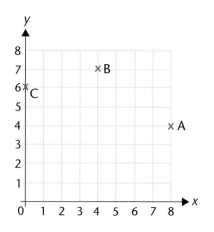

EXAMPLE 4

a) State the coordinates of the points marked D, E and F.

b) Copy the grid and mark the points G(4, 8), H(6, 11) and I(7, 15).

c) Find the coordinates of the point halfway along line DJ.

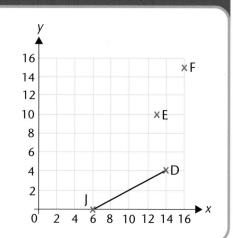

EXAMPLE 4

a) D is (14, 4), E is (13, 10) and F is (16, 15).
The lines are marked 2, 4, 6, … so 13 is halfway between 12 and 14.

b)

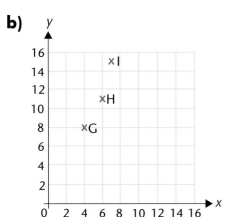

> **EXAM TIP**
> A common error when plotting points is to label the spaces instead of the lines.

c) To get from point D to point J you move 8 units to the left and 4 units down.

To get halfway along the line between D and J you move 4 units to the left and 2 units down.

The coordinates of the point halfway along the line are (10, 2).

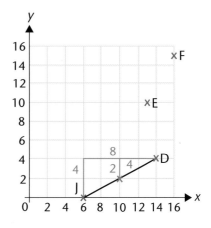

The point halfway along a line is called the **midpoint** of the line.

EXERCISE 3.2

1 Write down the coordinates of the points marked A, B and C.

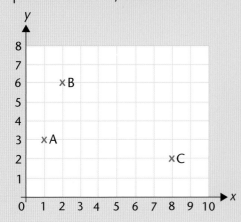

2 Write down the coordinates of the points marked A, B and C.

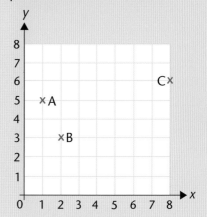

3 Write down the coordinates of the points marked D, E and F.

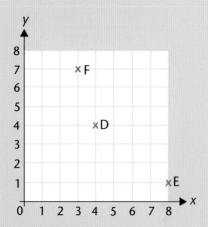

5 Write down the coordinates of the points marked L, M and N.

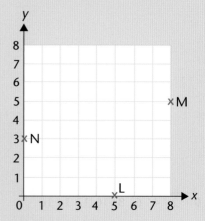

4 Write down the coordinates of the points marked D, E and F.

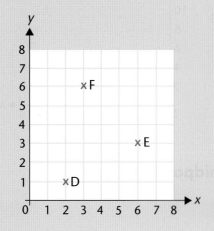

6 Write down the coordinates of the points marked G, H and I.

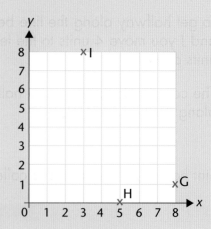

7 Write down the coordinates of the points marked P, Q and R.

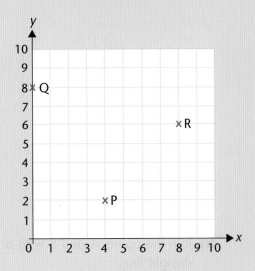

8 Write down the coordinates of the points marked P, Q and R.

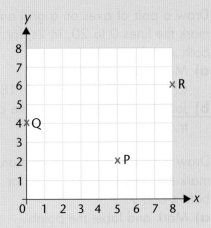

9 Draw a pair of axes on a grid and mark the lines 0 to 8 for both x and y (as in question **8**).
Mark and label the points A(4, 7), B(1, 3) and C(5, 8).

10 Draw a pair of axes on a grid and mark the lines 0 to 8 for both x and y.
Mark and label the points A(5, 1), B(2, 3) and C(4, 6).

11 Draw a pair of axes on a grid and mark the lines 0 to 8 for both x and y.
Mark and label the points S(5, 3), T(7, 2) and W(0, 5).

12 Draw a pair of axes on a grid and mark the lines 0 to 8 for both x and y.
Mark and label the points S(1, 3), T(8, 2) and W(4, 0).

13 Draw a pair of axes on a grid and mark the lines 0 to 8 for both x and y.
Mark and label the points M(2, 8), N(7, 7) and R(5, 0).

14 Draw a pair of axes on a grid and mark the lines 0 to 10 for both x and y.
Mark and label the points M(2, 5), N(4, 7), P(0, 8) and R(9, 10).

STAGE

1

15 Write down the coordinates of the points marked A, B, C and D.

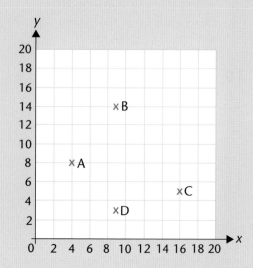

16 Write down the coordinates of the points marked A, B, C and D.

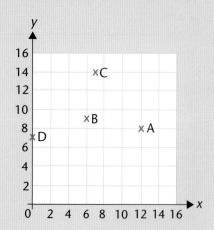

17 Draw a pair of axes on a grid and mark the lines 0 to 8 for both x and y.
   a) Mark and label the points A(1, 4) and B(5, 4).
   b) Join the points A and B with a straight line.
   c) Write down the coordinates of the midpoint of the line AB.

18 Draw a pair of axes on a grid and mark the lines 0 to 8 for both x and y.
   a) Mark and label the points A(3, 4) and B(3, 8).
   b) Join the points A and B with a straight line.
   c) Write down the coordinates of the midpoint of the line AB.

19 Draw a pair of axes on a grid and mark the lines 0 to 20, in 2s, for both x and y.
   a) Mark and label the points C(2, 18), D(2, 9) and E(15, 18).
   b) Join the three points to make a triangle.

20 Draw a pair of axes on a grid and mark the lines 0 to 20, in 2s, for both x and y.
   a) Mark and label the points A(5, 12), B(15, 6), C(4, 2) and D(16, 16).
   b) Join the points A and B and join the points C and D.
   c) Write down the coordinates of the point where the two lines cross.

A ACTIVITY 4

Cover up

This is a game for two students.

You need two dice of different colours, for example one red dice and one blue dice.

Draw a grid like the one here.

Take turns to roll the two dice.

The score on the blue dice will give the *x*-coordinate and the score on the red dice the *y*-coordinate of a point.

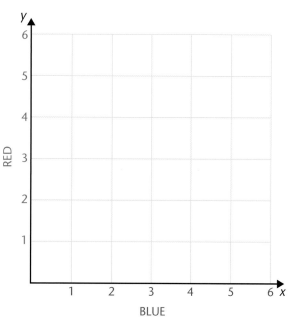

Put a cross at the point if you are Player 1 or a circle if you are Player 2.
If the point if already taken, play passes back to your partner.

Continue taking turns to throw the dice and mark points until all of the grid has been covered.

The player with the most crosses or circles wins.

K KEY IDEAS

- The eight main compass directions are north, north east, east, south east, south, south west, west, north west.

- You should know the name of the *x*-axis, the *y*-axis and the origin.

- You should be able to locate a point on a grid, given its coordinates.

- You should know how to write down the coordinates of a point (across, up) or (*x*, *y*).

STAGE
1

4 Scales

Telling the time

The two clock faces below show what the minute hand (the long hand) tells you when it points to the numbers on a clock.

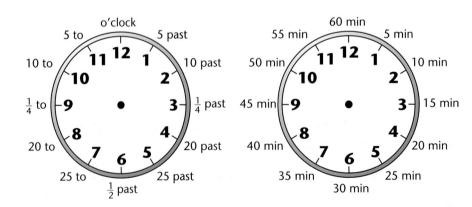

4

EXAMPLE 1

Write the time shown on each of these clocks in words and in figures.

a)

b)

a) five past six, 6:05

b) twenty-five to four, 3:35

EXERCISE 4.1

For each of questions **1** to **18**, write the time in words and in figures.

1

3

2

4

STAGE

1

EXERCISE 4.1 continued

5
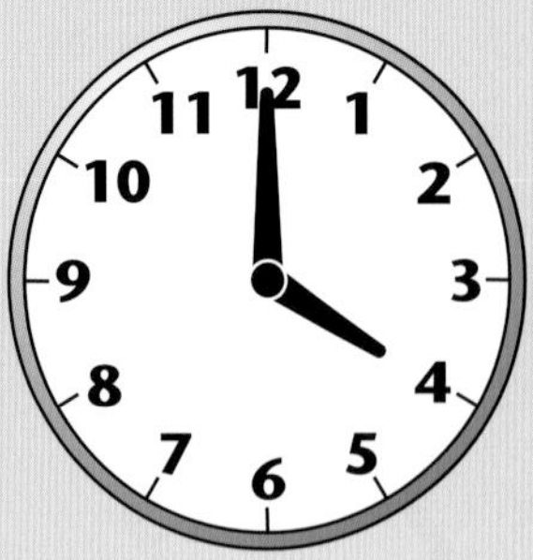

6
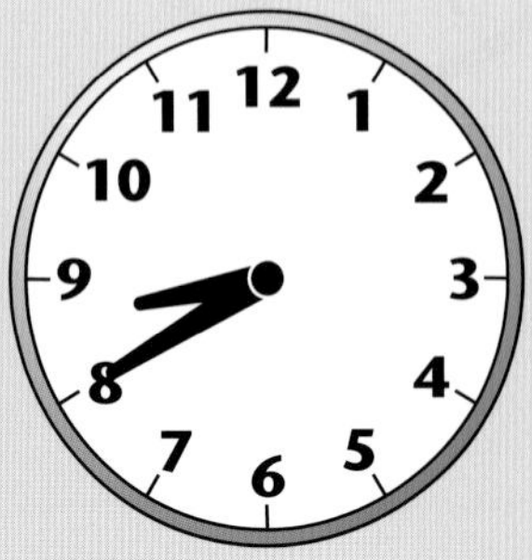

7
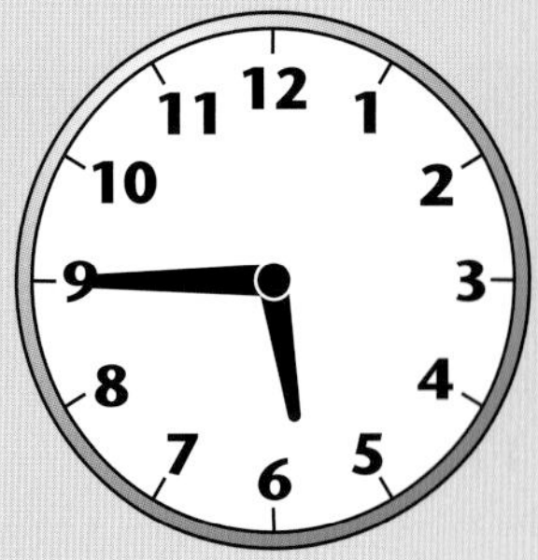

8
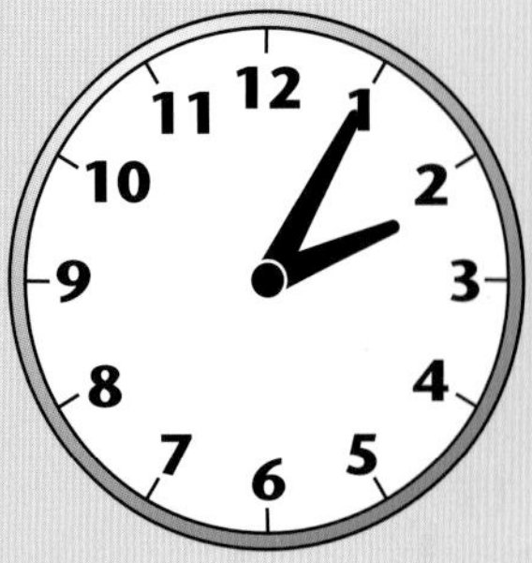

9
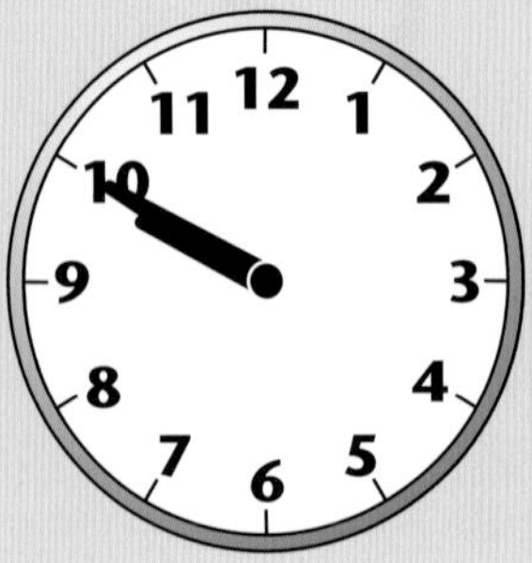

10
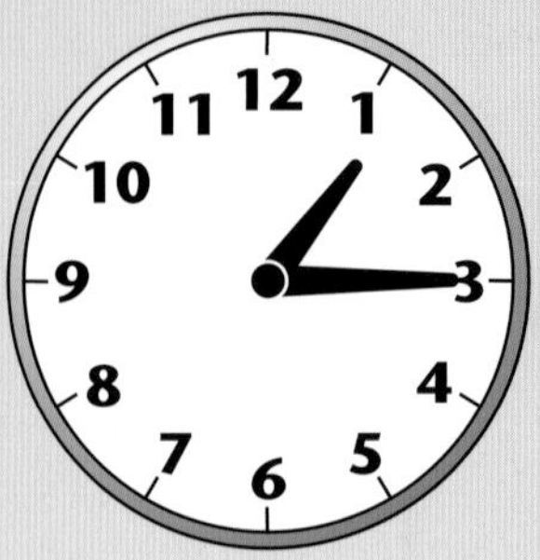

11
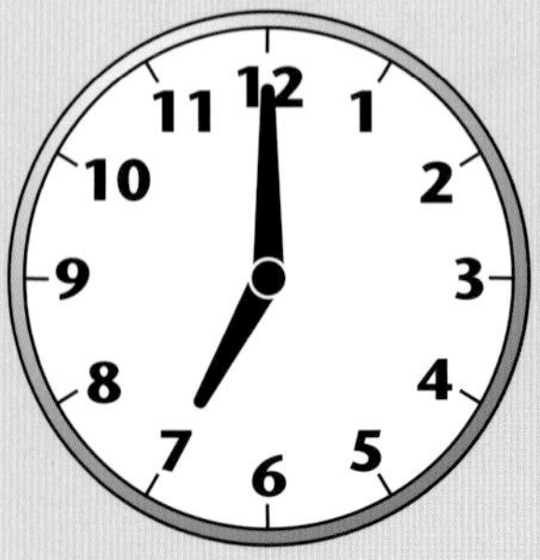

12
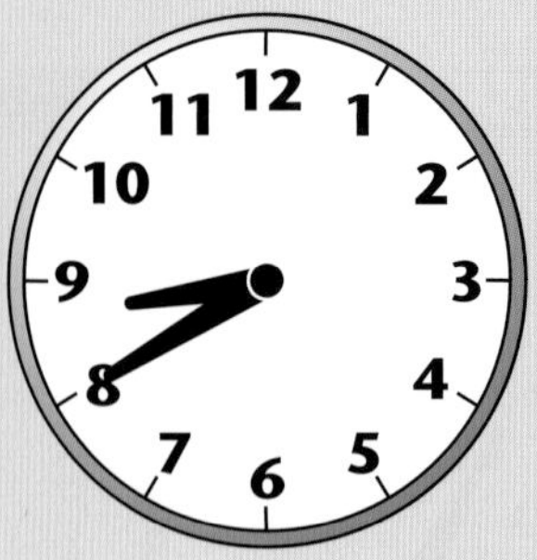

13
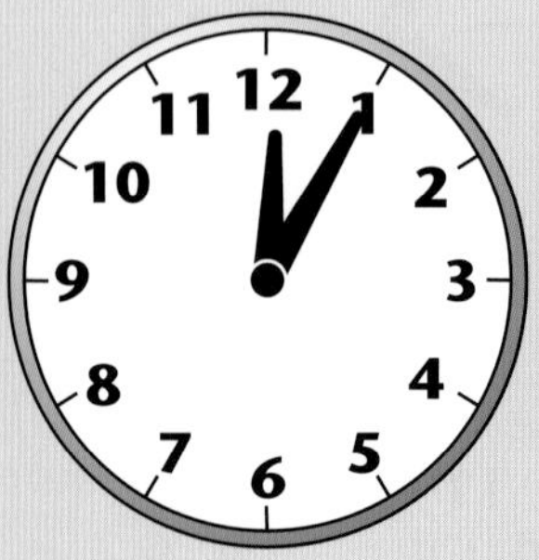

14
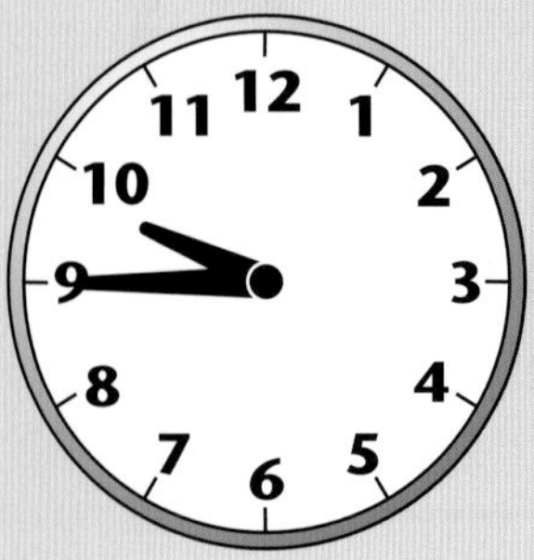

15

16

17

18

For each of questions **19** to **42**, draw a clock face and mark on it the time given.

19 10 past 6

20 $\frac{1}{4}$ past 2

21 5 o'clock

22 6:50

23 10:35

24 25 past 7

25 10 to 12

26 $\frac{1}{4}$ to 3

27 3:45

28 12:40

29 25 to 6

30 1:55

31 25 to 5

32 12 o'clock

33 $\frac{1}{2}$ past 9

34 11:20

35 7:45

36 20 past 2

37 $\frac{1}{4}$ to 4

38 6:55

39 $\frac{1}{2}$ past 5

40 11:05

41 10 to 8

42 9:50

Scales

4

STAGE

1

4

How long does it take?

EXAMPLE 2

How many minutes are there between these times?

a) 7:25 and 7:50

b) $\frac{1}{4}$ to 10 and 25 past 10

a) From 25 minutes to 50 minutes is 25 minutes.

b) From $\frac{1}{4}$ to 10 up to 10 o'clock is $\qquad$ 15 minutes

From 10 o'clock to 25 past 10 is $\qquad$ + 25 minutes

$\qquad$ 40 minutes

So from $\frac{1}{4}$ to 10 to 25 past 10 is 40 minutes.

EXERCISE 4.2

For each of questions **1** to **20**, find how many minutes there are between the times. You can use a clock face to help you.

	Start time	Finish time			Start time	Finish time
1	8:00	8:15		**11**	2:30	3:10
2	2:30	2:55		**12**	2:50	3:15
3	6:20	6:50		**13**	11:55	12:20
4	10:10	10:45		**14**	8:20	9:05
5	4:10	4:35		**15**	$\frac{1}{4}$ to 7	20 past 7
6	3:05	4:00		**16**	4:35	5:15
7	$\frac{1}{4}$ past 3	20 to 4		**17**	1:40	2:35
8	$\frac{1}{4}$ past 11	$\frac{1}{4}$ to 12		**18**	$\frac{1}{2}$ past 10	5 past 11
9	$\frac{1}{4}$ to 10	5 to 10		**19**	10 to 4	25 to 5
10	20 past 7	20 to 8		**20**	25 to 9	$\frac{1}{2}$ past 9

EXERCISE 4.2 continued

21 Emily's lunch break starts at 12 o'clock.
French Club starts at 12:20.
How long does she have to eat her lunch?

22 Roger goes canoeing.
He leaves the boathouse at 4:05 and returns at 4:55.
For how long is he canoeing?

23 Mrs Roberts leaves work at 10 past 6.
She reaches home at 10 to 7.
How long does her journey take?

24 Jack walks to school.
He leaves the house at $\frac{1}{4}$ to 9.
He gets to school just in time, at 5 to 9.
How long does it take him to walk to school?

25 Maria checks the clock as she enters the supermarket.

a) What time does the clock show?

She looks at the clock again as she leaves.

b) How long does she take to do her shopping?

26 A train leaves West Kirby station at $\frac{1}{4}$ to 6.
It arrives in Liverpool at $\frac{1}{4}$ past 7.
How long is the journey?

27 A TV programme starts at 7:30 and ends at 8:05.
How long is the programme?

28 Mark starts his session on the cross-trainer at the gym at 2:50.
He moves on to the rowing machine at 3:05.
For how long does he use the cross-trainer?

C CHALLENGE 1

How long is there between these times?

Give your answers in minutes and in hours and minutes.

	Start time	Finish time
a)	$\frac{1}{2}$ past 7	$\frac{1}{4}$ to 9
b)	8:35	11:40
c)	3:25	6:10
d)	10 to 4	25 to 8
e)	11:20	3:15

STAGE
1

Finishing times

EXAMPLE 3

Work out the finishing time for each of these.

a) Starts at 1:35 and lasts for 15 minutes.

b) Starts at $\frac{1}{2}$ past 9 and lasts for 45 minutes.

a) Add 15 on to 35 to give 50. The finishing time is 1:50.

b) Add 30 of the 45 minutes to take it up to 10 o'clock.
Add on the remaining 15 minutes, giving a finishing time of 10:15.

EXERCISE 4.3

1 Work out the finishing time for each of these.

	Starts	Lasts
a)	2:05	30 minutes
b)	4:15	50 minutes
c)	11:35	25 minutes
d)	5 to 9	45 minutes
e)	6:10	35 minutes
f)	8:55	20 minutes
g)	5:30	15 minutes
h)	20 to 1	15 minutes
i)	$\frac{1}{4}$ past 10	25 minutes

2 A plane leaves Heathrow at 10:05 and flies to Leeds.
The flight lasts 45 minutes.
At what time does the plane land?

3 My favourite TV programme lasts for 45 minutes. I start watching it at 7:05.
At what time does it finish?

4 Break time starts at 10:25 and lasts for 20 minutes.
At what time does school re-start?

5 A train leaves Chesterfield Station at 20 past 1. It arrives in Sheffield 20 minutes later.
At what time does it arrive?

6 The TV news starts at $\frac{1}{4}$ to 9 and lasts for 20 minutes.
At what time does the news finish?

7 A rugby match starts at 2:40.
The first half lasts for 35 minutes.
At what time does the first half finish?

8 A boy starts delivering newspapers at 6:55. He finishes 55 minutes later.
At what time does he finish?

9 My bus to work takes 15 minutes.
I catch the bus at 8:50.
What time will I arrive at work?

Reading scales

To read a scale, decide what each small division represents and read it from there.

EXAMPLE 4

What are the readings at A and B on each of the scales below?

a)

10   20   30

A   B

b)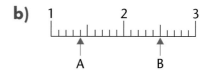

a) Here there are 10 divisions between 10 and 20 so each small division represents 1.

So A is 14 and B is 25.

b) Here there are 10 divisions between 1 and 2 so each small division represents 0·1.

So A is 1·4 and B is 2·5.

EXAMPLE 5

a) What is the reading at C on this scale?

50   60   70

C

b) Copy the scale and mark 63 with the letter D.

a) Here there are 5 divisions between 50 and 60, so each division represents 2.

So C is 56.

b) The divisions are at 62 and 64, so D is midway between.

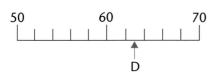

EXAMPLE 6

a) Read the point on the scale marked E.

b) Copy the scale and mark 650 with the letter F.

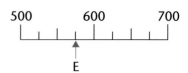

a) Here there are four divisions between 500 and 600, so each division represents 25.

So E is 575.

b)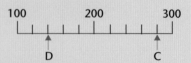

> ### EXAM TIP
> Most errors in reading scales are made because the smallest division is assumed to be 0·1, 1, 10, ... Make sure you know what the smallest division is before you start reading or marking on a scale.

EXERCISE 4.4

1 Read the points marked A and B on the scale below.

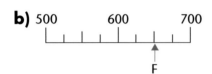

2 Read the points marked A and B on the scale below.

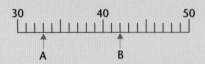

3 Read the points marked C and D on the scale below.

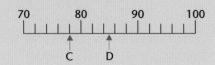

4 Read the points marked C and D on the scale below.

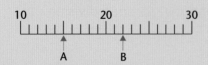

5 Read the points marked E and F on the scale below.

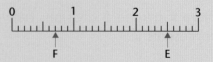

6 Read the points marked E and F on the scale below.

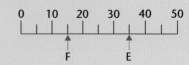

EXERCISE 4.4 continued

7 a) Read the point marked G on the scale below.
b) Copy the scale and mark H at 37.

8 Copy the scale below and mark G at 48 and H at 62.

9 a) Read the point marked I on the scale below.
b) Copy the scale and mark J at 250.

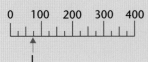

10 a) Read the point marked I on the scale below.
b) Copy the scale and mark J at 0·3.

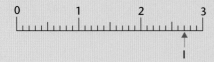

11 a) Read the point marked K on the scale below.
b) Copy the scale and mark L at 160.

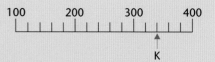

12 a) Read the point marked K on the scale below.
b) Copy the scale and mark L at 63.

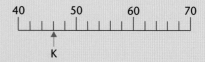

13 a) Read the point marked M on the scale below.
b) Copy the scale and mark N at 700.

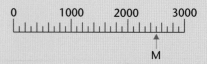

14 a) Read the points marked M and N on the scale below.
b) Copy the scale and mark Q at 105.

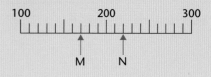

15 What is the reading on this thermometer?

16 What is the reading on this dial?

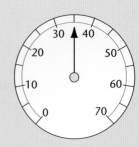

17 What is the reading on this dial?

18 What is the reading on this speedometer?

19 How much does this weigh?

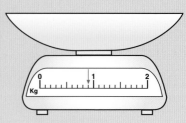

EXERCISE 4.4 continued

20 a) What is the reading on this dial?

b) Alan finds the accurate pressure from the internet.
It is 29·3.
Mark this on a copy of the diagram.

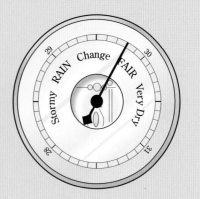

ACTIVITY 1

Design a dashboard for a car. Include

■ a speedometer going from 0 to 100 mph, marked in 5s.

■ a fuel gauge going from 0 to 1, marked in $\frac{1}{4}$s.

■ a temperature scale going from 0 to 30, marked in 1s.

KEY IDEAS

■ You should know how to tell the time and how to write times in words and in figures.

■ There are strategies for finding time intervals and finishing times.

■ There are strategies for reading scales with different graduations.

STAGE

1

Revision exercise B1

1 Look at this grid.

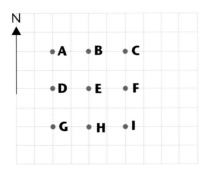

a) What is the direction
  (i) from A to B?
  (ii) from B to D?
  (iii) from H to E?
  (iv) from G to E?
  (v) from I to H?
b) Name the letter that is
  (i) north of E.
  (ii) south east of B.

2 John walked along a road that went south west.
David walked the other way along the same road.
In what direction did David walk?

3 Write the coordinates of each point marked on the grid below.

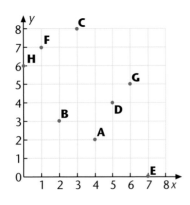

4 Draw axes like the ones in question **3**.
Plot and label each of these points.
I(1, 4), J(8, 2), K(0, 6), L(5, 7), M(2, 1), N(3, 5), P(4, 0), Q(7, 3)

5 Write the coordinates of each point marked on the grid below.

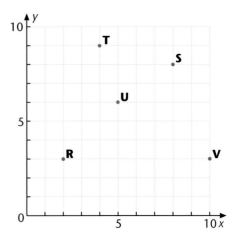

6 Copy this grid.
Plot and label each of these points.
W(1, 3), X(3, 7), Y(2, 5), Z(4, 8)

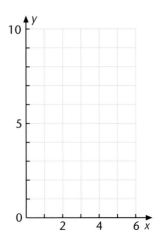

7 Write each of these times in words and in figures.

a)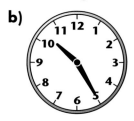

b)

c)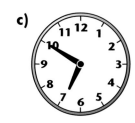

8 The late evening news starts at $\frac{1}{4}$ past 11 and lasts for 20 minutes.
At what time does it finish?

9 School finishes at 3:15 p.m. I arrive home at 4:05 p.m.
How long has it taken me to get home?

10 A train takes 35 minutes to travel from Nottingham to Leicester.
It leaves Nottingham at 2:45.
At what time does it arrive in Leicester?

11 Write down the points marked on each of these scales.

a)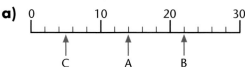

b)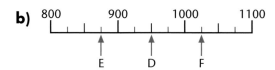

c)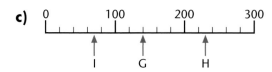

12 a) Copy this scale and mark on the points A 3·5, B 1·8 and C 2·4.

b) Copy this scale and mark on the points D 140, E 185, F 205 and G 162.

c) Copy this scale and mark on the points H 45, I 72 and J 89.

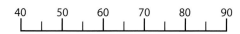

STAG
1

5 Algebra and patterns

You will learn about

- Using symbols for numbers
- Using letters for numbers
- Expressions in algebra
- Continuing number patterns
- Using function machines

You should already know

- How to add, subtract, multiply and divide whole numbers

Using symbols for numbers

Here is a number puzzle.

$$6 + \bullet = 14$$

What does the symbol ● stand for?

The totals on each side of the = sign must be the same.

To find what the symbol ● stands for you need to find the number that you must add to 6 to get 14.

The number is 8 because 6 + 8 = 14, so ● = 8.

Here is another number puzzle.

$$\blacksquare + \blacksquare = 14$$

What does the symbol ■ stand for?

Again, the totals on each side of the = sign must be the same.

This time there are two symbols but they are the same. This tells you that they stand for the same number.

Two times the number that the symbol ■ stands for is 14, so ■ = 7.

EXERCISE 5.1

In each question, find the number the symbol stands for.

1 $5 + \bullet = 12$

2 $15 + \clubsuit = 22$

3 $21 - \blacksquare = 12$

4 $42 - \blacklozenge = 30$

5 $\blacktriangle + 5 = 19$

6 $\bigstar + 17 = 30$

7 $\star - 7 = 20$

8 $\heartsuit - 23 = 5$

9 $8 + \clubsuit = 17$

10 $32 - \clubsuit = 26$

11 $3 + \odot = 3$

12 $\blacklozenge - 3 = 3$

13 $\flat + \flat = 30$

14 $* + * = 12$

15 $7 = 7 - \odot$

16 $12 - \diamond = 6 + \diamond$

17 $6 \times \blacksquare = 24$

18 $\blacktriangle \times 9 = 54$

19 $7 \times \heartsuit = 63$

20 $\blacklozenge \times \blacklozenge = 64$

21 $24 \div \bullet = 2$

22 $28 \div * = 4$

23 $\blacksquare \div 6 = 10$

24 $* \div 5 = 3$

Using letters for numbers

This line is made up of two pieces, 6 cm and 4 cm long.

You can see that the full length is 6 + 4 = 10 cm.

If the second piece is 8 cm the full length is 6 + 8 = 14 cm.

If the second piece is some length that is not known, a symbol can be used for the length. In this section, the symbols used are letters.

Here the length has been called x.

The full length is 6 + x or x + 6. This is called an **expression** for the length.

When a letter is used instead of a number it is called **algebra**.

Here the letter is x but any letter can be used.

STAGE
1

In the examples and exercises, the lines are not drawn accurately so do not measure them.

All lengths are in centimetres.

EXAMPLE 1

a) Write down the length of the full line.

b) Write down an expression for the length, in centimetres, of each of these lines.

(i)

(ii)

a) Length is 2 + 5 = 7 cm.

b) (i) Length is 2 + q.
   (ii) Length is 3 + y.

This line is x cm long altogether. One part is 3 cm, so an expression for the length, in centimetres, of the part marked ? is $x - 3$.

EXAMPLE 2

Write down an expression for each length marked ? on these lines.

a)

b)

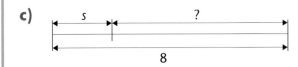

c)

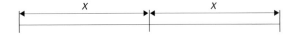

a) Length of ? is $x - 4$.

b) Length of ? is $y - 9$.

c) Length of ? is $8 - s$.

If two lines, each x cm long, are put end to end, the expression for the full length is $x + x$ or $2 \times x$, and this can be written as $2x$.

EXAMPLE 3

a) Write down an expression for the length of each of these lines.

(i)

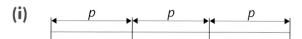

(ii)

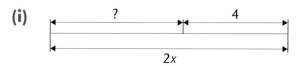

b) Write down an expression for each length marked ? on these lines.

(i)

(ii)

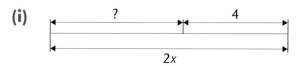

a) (i) Length is $p + p + p = 3p$.

(ii) Length is $s + s + 3 = 2s + 3$.

b) (i) Length of ? is $2x - 4$.

(ii) Length of ? is $8 - 2d$.

EXAM TIP

- $p + 5$, $t - 3$, $8 - x$ and other similar expressions cannot be written more simply, but $p + 3$ is the same as $3 + p$.
- $n + n$ or $2 \times n$ can be written as $2n$.

STAGE

1

EXERCISE 5.2

1 a) Write down the length of this line.

b) Write down an expression for the length of each of these lines.

(i)

(ii)

2 a) Write down the length of this line

b) Write down an expression for the length of each of these lines.

(i)

(ii)

3 a) Write down the length of this line.

b) Write down an expression for the length of each of these lines.

(i)
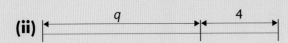

(ii)

▌▌▌ EXERCISE 5.2 continued

4 a) Write down the length of this line.

b) Write down an expression for the length of each of these lines.

(i)

(ii)

5 Write down an expression for the length of each of these lines.

a)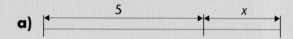

b)

c)

d)

6 a) Write down the length of this line.

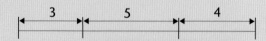

b) Write down an expression for the length of this line.

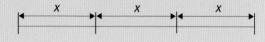

7 a) Write down the length of this line.

b) Write down an expression for the length of this line.

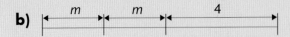

8 Write down an expression for the length of each of these lines.

a)

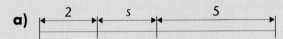

b)

c)

d)

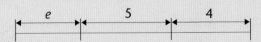

9 a) Write down the length marked ? on this line.

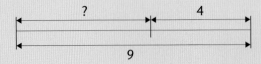

b) Write down an expression for the length marked ? on this line.

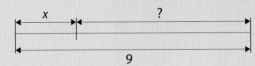

10 Write down an expression for the length of each of these lines.

a)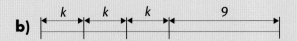

b)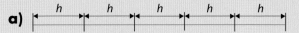

11 a) Write down the length marked ? on this line.

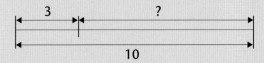

b) Write down an expression for the length marked ? on this line.

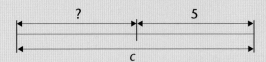

12 Write down an expression for each length marked ? on these lines.

a)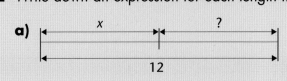

b)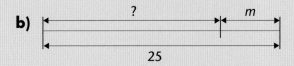

c)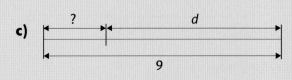

d)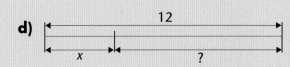

STAGE
1

Number patterns

You have already met some number patterns.

These are the counting numbers.

1   2   3   4   5   6   7   ...

These are the odd numbers.

1   3   5   7   9   11   13   ...

You know two names for this number pattern.

What are they?

2   4   6   8   10   12   14   ...

The times tables are number patterns too. This is the 3 times table.

3   6   9   12   15   18   21   ...

The next number in this number pattern is 24.

If you did not know your 3 times table you could find the next number by adding 3 to the last one.

EXAMPLE 4

Look at this number pattern.

7   14   21   28   ...

What is the next number in the pattern?

Explain how you found the next number.

This is the 7 times table so the next number is 35.
You can work out the next number by saying your 7 times table.

$1 \times 7 = 7, \quad 2 \times 7 = 14, \quad 3 \times 7 = 21, \quad 4 \times 7 = 28, \quad 5 \times 7 = 35$

Alternatively, you can add 7 to the last number.

$28 + 7 = 35$

In some patterns the numbers get smaller.

60   55   50   45   40   35   30   ...

You can see that the numbers are in the 5 times table but they are getting smaller instead of bigger.

To get to the next number in the pattern you subtract 5.

So the next number is

30 − 5 = 25.

Here is a pattern that is not one of the times tables.

3   7   11   15   19   23   27   ...

To get from one number to the next you add 4.

+ 4   + 4   + 4   + 4   + 4   + 4

3   7   11   15   19   23   27   ...

EXAMPLE 5

What is the next number in this pattern?

4   10   16   22   28   34   40   ...

To get from one number to the next you add 6.

+ 6   + 6   + 6   + 6   + 6   + 6   + 6

4   10   16   22   28   34   40   ...

So the next number is

40 + 6 = 46.

STAGE
1

Algebra and patterns

1 Look at this number pattern.

4   8   12   16   …

What are the next two numbers in this pattern?

2 Look at this number pattern.

0   2   4   6   …

What are the next two numbers in this pattern?

3 Look at this number pattern.

9   18   27   36   …

a) What is the next number in this pattern.
b) Explain how you worked out the next number in the pattern.

4 Look at this number pattern.

1   3   5   7   …

What are the next two numbers in this pattern?

5 Look at this number pattern.

1   5   9   13   …

What are the next two numbers in this pattern?

6 Look at this number pattern.

1   4   7   10   …

a) What are the next two numbers in this pattern?
b) Explain how you work out the next number in the pattern.

STAGE
1

⫴ EXERCISE 5.3 continued

7 Look at this number pattern.

26   23   20   17   ...

a) What are the next two numbers in this pattern?
b) Explain how you work out the next number in the pattern.

8 For each of these number patterns
  (i) write down the next two numbers in the pattern.
  (ii) explain how you work out the next number in the pattern.

a) 1   7   13   19   ...
b) 20   18   16   14   ...
c) 100   101   102   103   ...
d) 5   13   21   29   ...
e) 77   67   57   47

Function machines

A function machine consists of an input, a rule and an output.

In the function machine below, the rule adds 3 to the input of 6 to give an output of 9.

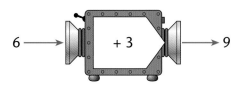

STAGE
1

EXERCISE 5.4

Find the outputs from each of these function machines.

1
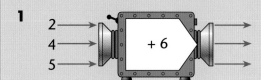

2

3
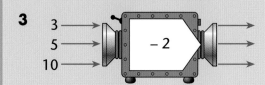

4
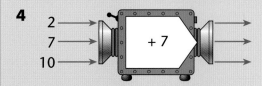

5
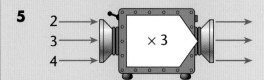

6
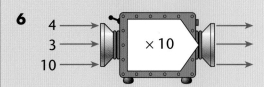

7

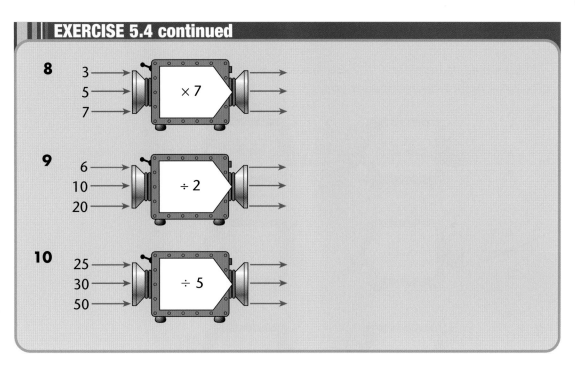

8 3 → ×7 →
     5 →   →
     7 →   →

9 6 → ÷2 →
     10 →   →
     20 →   →

10 25 → ÷5 →
      30 →   →
      50 →   →

A function machine can have two or more rules in a row.

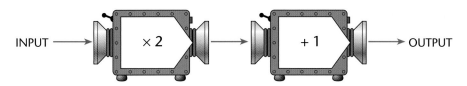

INPUT → ×2 → +1 → OUTPUT

With these inputs you will get the following outputs.

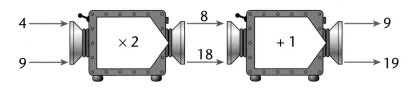

4 → ×2 → 8 → +1 → 9
9 →   → 18 →   → 19

STAGE
1

EXERCISE 5.5

Find the outputs from each of these function machines.
The first one has been done for you.

1

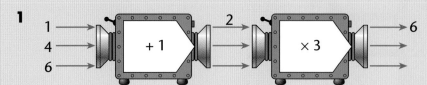

1 → [+ 1] → 2 → [× 3] → 6
4 →
6 →

2

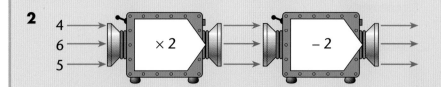

4 → [× 2] → [− 2] →
6 →
5 →

3

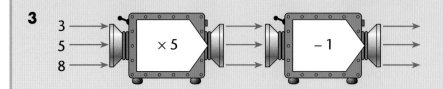

3 → [× 5] → [− 1] →
5 →
8 →

4

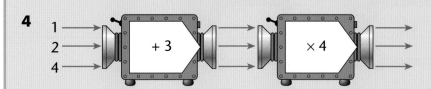

1 → [+ 3] → [× 4] →
2 →
4 →

5

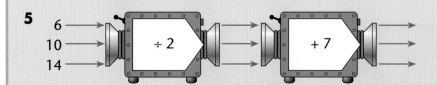

6 → [÷ 2] → [+ 7] →
10 →
14 →

6

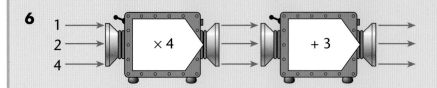

1 → [× 4] → [+ 3] →
2 →
4 →

7

1 → [× 4] → [× 2] →
5 →
10 →

8

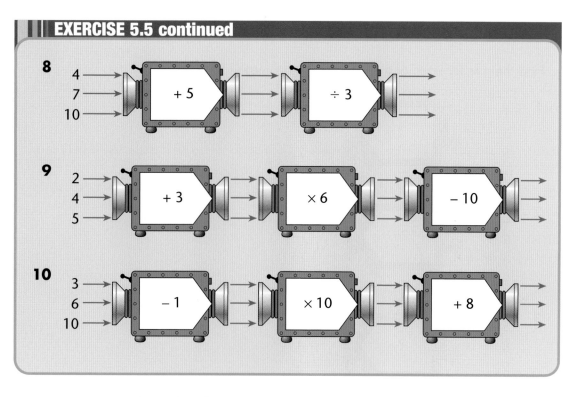

9

10

Sometimes you will need to find the rule for the function machine.

Here there are two possible rules, × 2 and + 2.

Sometimes there is more than one input and output pair. If so, then the rule must work for all of them. So for the function machine below, the rule must be × 2.

STAGE

1

67

EXERCISE 5.6

Find the rule for each of these function machines.

1

1 → ? → 3
3 → ? → 5
5 → ? → 7

2

2 → ? → 12
4 → ? → 14
5 → ? → 15

3

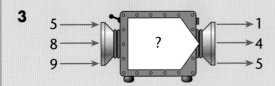

5 → ? → 1
8 → ? → 4
9 → ? → 5

4

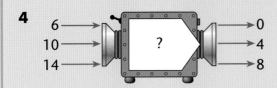

6 → ? → 0
10 → ? → 4
14 → ? → 8

5

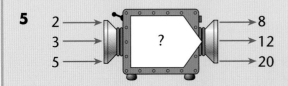

2 → ? → 8
3 → ? → 12
5 → ? → 20

6

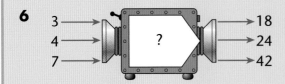

3 → ? → 18
4 → ? → 24
7 → ? → 42

7

16 → ? → 2
24 → ? → 3
40 → ? → 5

EXERCISE 5.6 continued

8

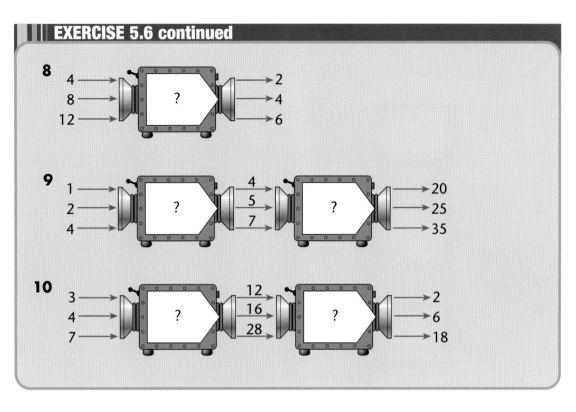

9

10

When you know the rule and the output, you can find the input by working backwards and using the inverse rule.

The inverse function machine will be

which will give an answer of 5.

EXAM TIP

You can sometimes work these out in your head but it is always worth showing your working.

STAGE

1

EXERCISE 5.7

Find the input values for each of these function machines.

1

2
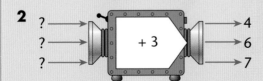

3

4
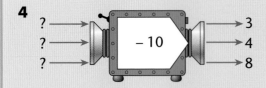

5
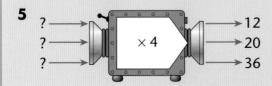

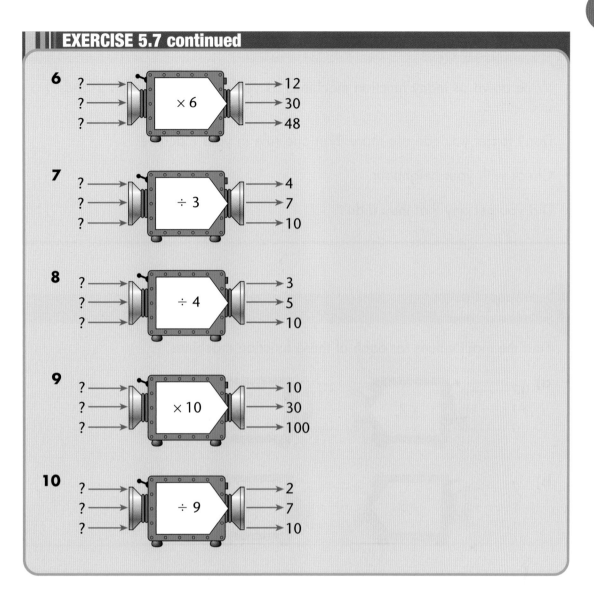

6
? → ×6 → 12
? → ×6 → 30
? → ×6 → 48

7
? → ÷3 → 4
? → ÷3 → 7
? → ÷3 → 10

8
? → ÷4 → 3
? → ÷4 → 5
? → ÷4 → 10

9
? → ×10 → 10
? → ×10 → 30
? → ×10 → 100

10
? → ÷9 → 2
? → ÷9 → 7
? → ÷9 → 10

A ACTIVITY 1

Write down as many function machines as you can that start with 2 and end with 6.

Don't forget you can use more than one rule in your machine.

Check with your neighbour.

Did you get any that they didn't?

C CHALLENGE 1

Find the input values for each of these function machines.

a)

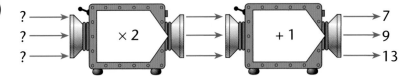

? → × 2 → + 1 → 7
? → → 9
? → → 13

b)

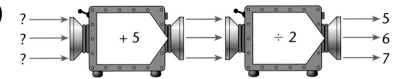

? → + 5 → ÷ 2 → 5
? → → 6
? → → 7

- Symbols can be used to stand for numbers.

- The symbols can be letters.

- An expression is a combination of letters and numbers.

- Letters added to pure numbers cannot be combined into a single expression, e.g. 5 plus *a* cannot be written more simply than $5 + a$.

- Letters subtracted from pure numbers cannot be combined into a single expression, e.g. 8 minus *b* cannot be written more simply than $8 - b$.

- Letters and numbers can be multiplied to give a single expression, e.g. *k* multiplied by 4 can be written as $4k$.

- Letters and numbers can be divided to give a single expression,

  e.g. *m* divided by 2 can be written as $\frac{m}{2}$.

- Letters which are the same can be combined, e.g. $x + x = 2x$.

- In a number pattern the numbers increase or decrease by the same amount each time.

- A function machine consists of an input, a rule and an output.

- When finding a rule, it must work for all input/output pairs.

- When finding the input, the inverse rule can be used.

STAGE

1

6 Problems and fractions

You will learn about

- Solving problems
- Interpreting a calculator display
- Using fractions

You should already know

- How to add, subtract, multiply and divide whole numbers without a calculator

Solving problems

When you are asked to solve a problem you have to identify the calculation that you need to do.

EXERCISE 6.1

1 Don has £69 in his savings account. He adds £25 more. How much does he have now?

2 A chocolate bar costs 35p and a packet of crisps costs 28p. How much do they cost altogether?

3 Sophie has 41 stamps. Her brother has 32 more than her. How many stamps does Sophie's brother have?

4 There are 17 people on a bus. At the next stop another 18 get on.
How many people are there on the bus now?

5 There are three Year 10 classes in school.
There are 24 students in one class, 28 in another and 31 in the third class.
How many students are there altogether in the three classes?

6 Rebecca collects £58 for charity.
Emma collects £33.
How much is this in total?

7 A box contains 35 chocolates.
8 are eaten.
How many are left?

8 Mum makes 78 sandwiches for
a party. 53 are eaten.
How many are left?

9 A piece of wood is 89 cm long.
Anne cuts off 63 cm to finish her
bookcase.
What length of wood is left?

10 Gran is 80 years old and
Elaine is 16.
How many years older than Elaine
is Gran?

11 I buy a CD costing £12.
I pay for it with a £50 note.
How much change will I get?

12 Two numbers added together
equal 42.
One of the numbers is 29.
What is the other number?

13 A box of chocolates costs £5.
How much will 9 boxes cost?

14 How many days are there in
4 weeks?

15 There are 7 rows of chairs in the
church hall.
Each row contains 8 chairs.
How many chairs are there
altogether?

16 There are 12 eggs in a box.
How many eggs are there in
5 boxes?

17 The school minibus has 14 seats.
How many seats are there on
3 such minibuses?

18 A lollipop costs 8p.
How much will 6 lollipops cost?

19 A farmer collects 42 eggs from
his hens.
He puts them into boxes which will
each hold 6 eggs.
How many boxes will he fill?

20 In a PE lesson 24 students are
divided into 3 teams.
How many students are there in
each team?

21 Three friends win £27 in a raffle.
They share the money equally.
How much does each receive?

22 How many 20p stamps can I buy
for £1?

23 A piece of ribbon 1 metre long is
cut into pieces 10 cm long.
How many pieces will there be?

24 A lottery winner gives half of his
£38 winnings to charity.
How much is this?

Problems and fractions

STAGE
1

75

EXAMPLE 1

Chocolate bars cost 38p each.
I buy 35 chocolate bars and pay with a £20 note.
How much change should I get?

Using a calculator, 38 × 35 = 1330.
This amount is in pence and needs to be changed into pounds.
There are 100 pence in £1, so the amount in pounds is
1330 ÷ 100 = 13·3 = £13·30.
Change = £20·00 − £13·30 = £6·70.

You could also solve this problem by
changing 38p into pounds first.
38p = £0·38
Using a calculator,
0·38 × 35 = 13·3 = £13·30.
Change = £20·00 − £13·30 = £6·70.

EXAM TIP

Notice that the zero in the second
decimal place is not shown on the
calculator display. However, for money
we always write two digits after the
decimal point. We write 13·3 as £13·30
and 6·7 as £6·70.

EXERCISE 6.2

1 A shirt costs £33·75 and a tie costs
  £15·95.
  How much do they cost altogether?
  How much change will I get from
  £50?

2 A firm makes 2463 Easter eggs on
  one day and 1895 eggs on the
  next day. How many Easter eggs
  have they made altogether?

3 A sack of potatoes weighs 55 kg.
  How much will 17 of these sacks
  weigh?

4 How many 6 m lengths of wire can
  be cut from a piece 228 m long?

5 There are 684 bars of chocolate to
  be packed into boxes.
  36 chocolate bars are packed into
  each box.
  How many boxes will be needed?

6 Christmas dinner costs £18·70 per
  person. How much will it cost 4
  people to have Christmas dinner?

7 A holiday costs £489.
  How much will it cost 5 friends to
  go on this holiday?

8 At the beginning of the year a car
  had travelled 10 703 miles.
  At the end of the year it had
  travelled 19 241 miles.
  How many miles did it travel
  during the year?

9 A maths book costs £17·95.
  I pay for it with a £20 note.
  How much change will I get?

10 What is the total cost of 16 stamps
   at 32p each? How much will I get
   from £6?

Fractions

Fraction means 'part'. So a fraction can tell you how much of a shape is shaded in.

> The bottom number in a fraction is called the **denominator**.
> This tells you how many equal parts there are.

> The top number in a fraction is called the **numerator**.
> This tells you howmany of the equal parts you have or require.

EXAMPLE 2

Ben says that half of this shape is shaded.
Explain why he is wrong.

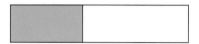

The shape is divided into two parts but the parts are not equal.

EXAMPLE 3

What fraction of this shape is

a) blue? **b)** not blue?

a) Since 2 of the 5 equal parts are blue, $\frac{2}{5}$ is blue.

b) Since 3 of the 5 parts are not blue, $\frac{3}{5}$ is not blue.

EXAMPLE 4

What fraction of this shape is blue?

Since 4 of the 16 equal parts are blue, $\frac{4}{16}$ is blue.

You can also see that the square can be divided into 4 larger squares.

So you could say that 1 of the 4 equal parts is blue so $\frac{1}{4}$ of the shape is blue.

6

STAGE
1

EXERCISE 6.3

What fraction of each shape is
a) blue?
b) not blue?

1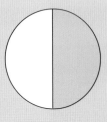

2

3

4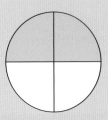

5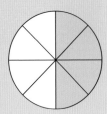

6

7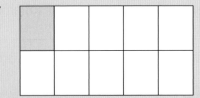

8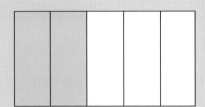

9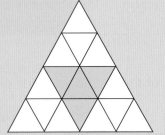

10

STAGE
1

EXERCISE 6.4

Copy each shape and shade the fraction required.

1 $\frac{3}{4}$

2 $\frac{1}{2}$

3 $\frac{1}{2}$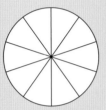

4 $\frac{1}{10}$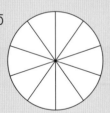

5 $\frac{1}{4}$

6 $\frac{4}{5}$

7 $\frac{1}{2}$

8 $\frac{1}{3}$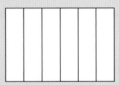

9 $\frac{4}{5}$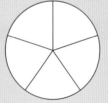

10 $\frac{1}{4}$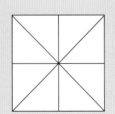

STAGE

1

Finding $\frac{1}{2}$, $\frac{1}{4}$ or $\frac{3}{4}$ of an amount

When finding $\frac{1}{2}$ of something you divide it by 2.

That is, you split it into two equal parts.

EXAMPLE 5

a) Find $\frac{1}{2}$ of 10.

b) Find $\frac{1}{2}$ of 3.

a) $10 \div 2 = 5$ and $5 + 5 = 10$.
Answer: $\frac{1}{2}$ of $10 = 5$.

b) $3 \div 2 = 1\frac{1}{2}$ and $1\frac{1}{2} + 1\frac{1}{2} = 3$.
Answer: $\frac{1}{2}$ of $3 = 1\frac{1}{2}$.

When finding $\frac{1}{4}$ of something you divide it by 4.

To do this you can halve it and halve it again.

EXAMPLE 6

a) Find $\frac{1}{4}$ of 20.

b) Find $\frac{1}{4}$ of 6.

a) $\frac{1}{2}$ of $20 = 10$ and $\frac{1}{2}$ of $10 = 5$.
Answer: $\frac{1}{4}$ of $20 = 5$.

b) $\frac{1}{2}$ of $6 = 3$ and $\frac{1}{2}$ of $3 = 1\frac{1}{2}$.
Answer: $\frac{1}{4}$ of $6 = 1\frac{1}{2}$.

When finding $\frac{3}{4}$ of something, you can find $\frac{1}{2}$ of it and $\frac{1}{4}$ of it and add the two answers together.

STAGE
1

EXAMPLE 7

Find $\frac{3}{4}$ of 80.

$\frac{1}{2}$ of 80 = 40

$\frac{1}{4}$ of 80 = 20   [$\frac{1}{2}$ of 80 = 40 and $\frac{1}{2}$ of 40 = 20]

Answer: $\frac{3}{4}$ of 80 = 40 + 20 = 60.

EXERCISE 6.5

1 Find $\frac{1}{2}$ of each of these.

a) 100   **b)** 30
c) 12   **d)** 15
e) 27   **f)** 20
g) 16   **h)** 70
i) 13   **j)** 25

2 Find $\frac{1}{4}$ of each of these.

a) 24   **b)** 80
c) 16   **d)** 18
e) 50   **f)** 40
g) 28   **h)** 100
i) 22   **j)** 30

3 Find $\frac{3}{4}$ of each of these.

a) 20   **b)** 60
c) 16   **d)** 10
e) 14   **f)** 6
g) 12   **h)** 80
i) 18   **j)** 30

C CHALLENGE 1

See if you can solve these problems.

a) $\frac{1}{2}$ of ? = 25

b) $\frac{1}{4}$ of ? = 14

c) $\frac{3}{4}$ of ? = 45

Can you explain to your partner how you found your answer?

Draw a poster to explain your method.

STAGE
1

Problems and fractions

K **KEY IDEAS**

- When you are asked to solve a problem you have to identify the calculation that you need to do.

- When you use a calculator you need to interpret the answer to make sure you answer the question. For example, an answer to a money problem needs two digits after the decimal point.

- The parts of a fraction are called the 'numerator' and the 'denominator'. The denominator tells you how many equal parts there are. The numerator tells you how many of the equal parts you have or require.

- To find $\frac{1}{2}$ of a quantity, you divide by 2.

- To find $\frac{1}{4}$ of a quantity, you divide by 4.

- To find $\frac{3}{4}$ of a quantity, you find $\frac{1}{2}$ and $\frac{1}{4}$ of the quantity and add them together.

STAGE
1

Revision exercise C1

1 For each part, write down the number that the symbol stands for.
   a) 15 + ★ = 23
   b) ● – 17 = 4
   c) 30 – ♦ = 21
   d) ■ + ■ = 28
   e) 4 × ♥ = 20
   f) ▲ ÷ 6 = 8

2 Write an expression for the length of each of these lines.

   a)
   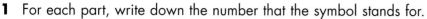

   b)

   c)

3 Write an expression for the length marked ? on each of these lines.

   a)

   b)

   c)

Revision exercise C1

4 For each of these number patterns
    (i) write down the next two terms in the pattern.
    (ii) explain how you work out the next number in the pattern.
a) 8 16 24   32   ...
b) 1 3 5   7   ...
c) 25 22 19   16   ...

5 Find the outputs from each of these function machines.

a)

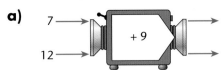

b)

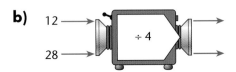

c)

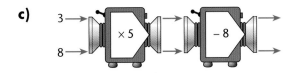

6 Find the rule for this function machine.

7 Find the input values for this function machine.

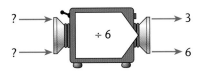

8 a) Beverley saves £3·50 a week for 6 weeks.
How much does she save altogether?

b) There are 41 people on a bus.
At the next stop 16 get off and 38 get on.
How many people are on a bus now?

c) Sarah buys a magazine costing £1·85.
Afterwards she has 65p left in her purse.
How much did she have before she bought the magazine?

d) Five people share £120 that they won on the National Lottery.
How much do they each get?

e) At the Steelers' last three games the attendances were 5831, 2387 and 1579.
What was the total attendance for all three games?

9 a) What fraction of the circle is
   (i) blue?
   (ii) not blue?

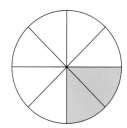

b) Copy this shape. Shade $\frac{4}{7}$ of it.

10 a) Find $\frac{1}{2}$ of each of these.
   (i) 16   **(ii)** 31   **(iii)** 164

b) Find $\frac{1}{4}$ of each of these.
   (i) 60   **(ii)** 26   **(iii)** 180

c) Find $\frac{3}{4}$ of each of these.
   (i) 36   **(ii)** 84   **(iii)** 50

STAG
1

Shapes

You will learn about

- Circles
- Special triangles and quadrilaterals
- Polygons
- Constructing regular polygons in a circle
- Enlargements

You should already know

- How to use ruler and compasses
- How to read scales

Circles

The distance all the way round a circle is its **circumference**.

A part of the circumference is called an **arc**.

The **radius** of a circle is the distance from the **centre** of the circle to its edge.

A line all the way across the circle and passing through its centre is a **diameter** of the circle.

So the length of the diameter = 2 × the length of the radius.

STAGE
1

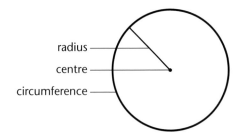
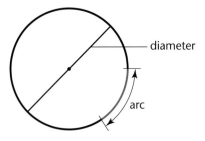

A ACTIVITY 1

Look around your classroom.

How many circles can you see?

Which object has the largest diameter?

Which object has the smallest radius?

Triangles

These shapes are all triangles. You can sort them into different groups.

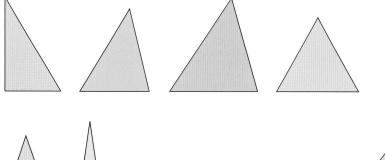

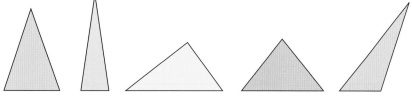

These are **right-angled triangles**.

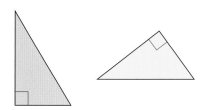

STAGE
1

Shapes

These are **isosceles triangles**.

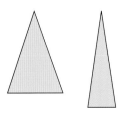

They have two sides the same length.

You can show the equal sides by marking them with a line like this.

These are **equilateral triangles**.

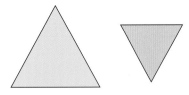

All their sides are the same length.
All their angles are the same size.

These triangles are **scalene triangles**.

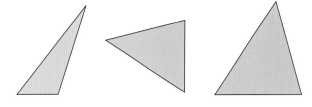

In scalene triangles all the sides and all the angles are different.

Quadrilaterals

Four-sided shapes are called **quadrilaterals**.

You know the **square** and the **rectangle**.

 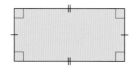

This is also a quadrilateral.

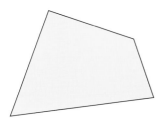

Polygons

A **polygon** is a many-sided shape.

Here are some common ones, apart from the triangles and quadrilaterals you have met already.

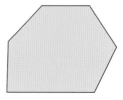

A **pentagon** has five sides.

A **hexagon** has six sides.

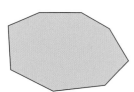

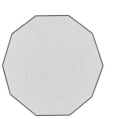

An **octagon** has eight sides.

A **decagon** has ten sides.

STAGE
1

A line joining two corners of a polygon is called a **diagonal**.

diagonal

When the sides of a polygon are all the same and the angles of the polygon are all the same, it is called a **regular polygon**.

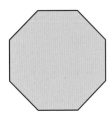 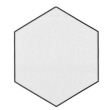 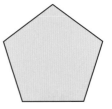

EXERCISE 7.1

1 Name each of these parts of a circle.

a)

b)

c)

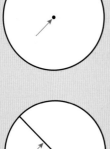

2 Name each of these polygons.

a)

b)

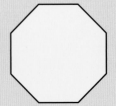

c)

EXERCISE 7.1 continued

3 Draw any pentagon.
Draw diagonals across the
pentagon from each vertex (corner)
of the pentagon to another vertex.
How many diagonals can you draw
in the pentagon altogether?

4 Draw any octagon.
Draw diagonals across the octagon
from each vertex of the octagon to
another vertex.
How many diagonals can you draw
in the octagon altogether?

Constructing a regular polygon

You will need a pair of compasses and a protractor or angle measurer.

Before you can draw any of the regular polygons you need to find the angle at the centre of the shape.

There are six equal angles at the centre of a regular hexagon.

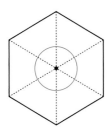

Since one complete turn is 360°, each of these angles will be
360 ÷ 6 = 60°.

Now you are ready to draw the shape.

1 Mark a centre.
Draw a circle with a radius of 4 cm
from the centre.

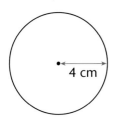

4 cm

STAGE

1

2 Mark a point at the top of the circle.
Measure an angle of 60° at the centre and mark a second point.

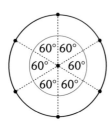

3 Choose method **a)** or **b)**.
a) Continue measuring angles of 60° and mark points all around the circle.

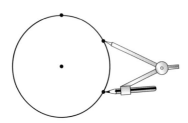

or

b) Open your compasses to the gap between the two points.
Mark off this gap all around the circle until you get back to the top.

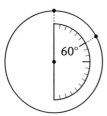

4 Join the points to make the regular hexagon.
You could rub out the circle if you wanted to.

EXERCISE 7.2

1 Construct a regular pentagon. (5 sides and 5 angles at the centre. Work out 360 ÷ 5.)

2 Construct a regular octagon (8 sides).

3 Construct a regular decagon (10 sides).

4 Construct a regular quadrilateral (4 sides – a square).

5 Construct a regular triangle (3 sides – an equilateral triangle).

6 Construct a regular nonagon (9 sides).

ACTIVITY 2

Construct a square.

Join the midpoints of the sides to form a smaller square inside the first square.

Continue to do this to make a pattern.

Colour in your pattern.

Remember:
The midpoint of a line is halfway along the line.

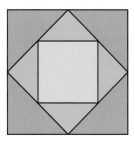

STAGE
1

CHALLENGE 1

Draw the same type of pattern as in Activity 2 for a pentagon or an octagon.

A · ACTIVITY 3

Mark the points for a regular hexagon around a circle.

Join them to make this shape.

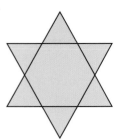

C · CHALLENGE 2

Try to make other shapes by using the points for other regular polygons.

Enlargements

Every length in shape B is two times as long as the length in shape A.

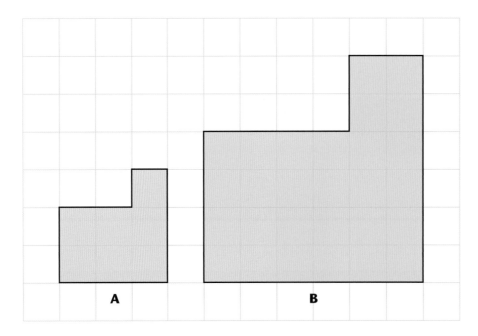

A

B

We say that shape B is a two times **enlargement** of shape A.

EXERCISE 7.3

1 Draw an enlargement of each of these on centimetre-square paper.
Make a two times enlargement or a three times enlargement as stated.

a) Two times

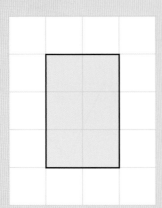

b) Three times

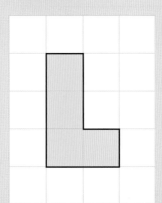

c) Two times

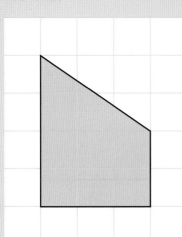

d) Two times

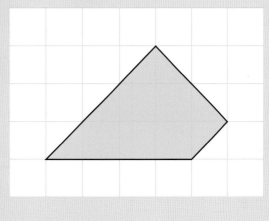

e) Two times

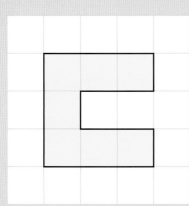

f) Three times

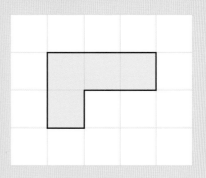

EXERCISE 7.3 continued

g) Two times

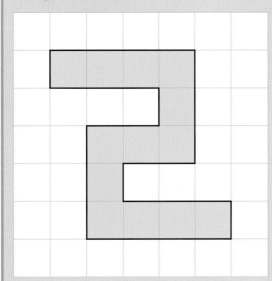

h) Two times

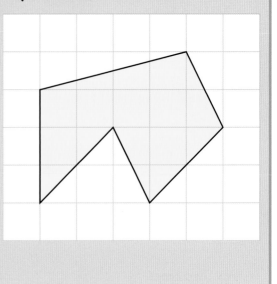

2 Measure the lengths in each pair of diagrams to see if one shape is an enlargement of the other.

a)

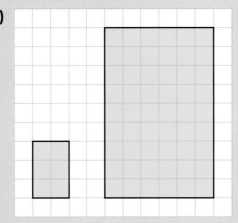

b)

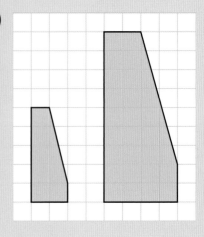

EXERCISE 7.3 continued

c)

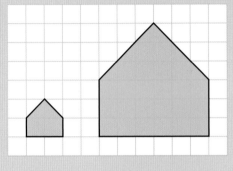

d)

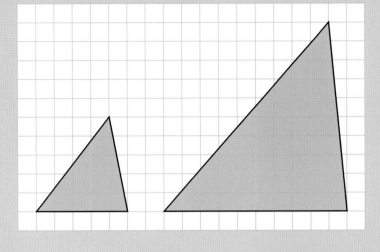

 ACTIVITY 4

On centimetre-square paper draw an everyday shape like the one below.
Then draw an enlargement of your shape.
Choose how many times bigger you want to make the lengths.

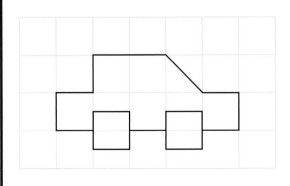

7

K KEY IDEAS

- The distance all the way round a circle is its circumference.

- The radius of a circle is the distance from the centre of the circle to its edge.

- A line all the way across a circle and passing through its centre is a diameter of the circle.

- A right-angled triangle has one angle equal to 90°.

- An isosceles triangle has two sides equal.

- An equilateral triangle has all sides of equal length and all angles equal.

- A scalene triangle has all sides different and all angles different.

- A quadrilateral has four sides.

- Squares and rectangles are special types of quadrilateral.

- A polygon is a many-sided shape.

- A pentagon has five sides.

- A hexagon has six sides.

- An octagon has eight sides.

- A regular polygon has all sides of equal length and all angles equal.

- You can construct a regular polygon in a circle using compasses and a protractor or angle measurer.

- In an enlargement, every length is the same number of times bigger than the original length.

Units, perimeter, area and volume

8

You will learn about

- Units of length
- Perimeter and its units
- Area and its units
- Volume and its units

You should already know

- How to multiply and divide by 10

Length

Lengths can be measured either in **imperial** (old) units – inches, feet, yards and miles – or in **metric** units – millimetres, centimetres, metres and kilometres.

You may have used feet and inches to measure your height, and miles to measure distance.

 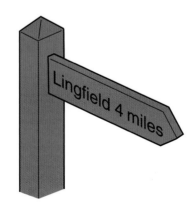

Height = 5 feet 3 inches

You may not have used yards at all as they are not used very often now.

Much more important now is the metric system.

You may use millimetres or centimetres to measure your handspan, metres or centimetres to measure your height, and kilometres to measure a long distance.

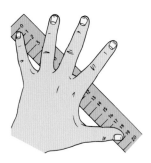

Handspan = 190 mm or 19 cm

Height = 1·60 m or 160 cm

You do not need to write out the words in full each time.

These are the accepted short versions.

metre = m
millimetre = mm
centimetre = cm
kilometre = km

STAGE
1

EXAMPLE 1

What metric unit would you use to measure these?

a) The width of a desk

b) The length of the school playing field

c) The distance from Birmingham to London

a) Millimetres or centimetres

b) Metres

c) Kilometres

EXAM TIP
Most short lengths can be given in either centimetres or millimetres.

ACTIVITY 1

Look at these numbers.

21·375   145   0·003 24   64·8   7·0132   789·2

Use a calculator to do each of these calculations.

1 Multiply each of the numbers by 10. What do you notice?

2 Multiply each of the numbers by 100. What do you notice?

3 Multiply each of the numbers by 1000. What do you notice?

4 Divide each of the numbers by 10. What do you notice?

5 Divide each of the numbers by 100. What do you notice?

6 Divide each of the numbers by 1000. What do you notice?

Try to write a summary of what happens in each case.

STAGE
1

These are the connections between the metric units of length.

1 kilometre = 1000 metres	1 km = 1000 m
1 metre = 1000 millimetres	1 m = 1000 mm
1 metre = 100 centimetres	1 m = 100 cm
1 centimetre = 10 millimetres	1 cm = 10 mm

These connections are important and you need to be able to change between the units.

Because the connections are all multiples of 10, changing from one unit to another never changes the digits, it only changes their place value.

EXAM TIP

When changing from one metric unit to another, decide what to multiply or divide by and then move the digits by the corresponding number of places. Whole numbers can have a decimal point after the number and when dividing it is helpful to put in the decimal point.

EXAMPLE 2

Change each of these lengths in metres to millimetres.

a) 2·435 m

b) 3·52 m

c) 4 m

d) 3·05 m

As there are 1000 millimetres in a metre, to change units you just multiply by 1000.

This means that you need to move the digits three places to the left. If the number is a whole number, you need to add three zeros as placeholders.

a) $2·435 \times 1000 = 2435$ mm   The digits have moved 3 places.

b) $3·52 \times 1000 = 3520$ mm   To move the digits 3 places you need to add a zero.

c) $4 \times 1000 = 4000$ mm   A whole number, so add 3 zeros.

d) $3·05 \times 1000 = 3050$ mm   Again 1 extra zero is needed.

EXAMPLE 3

Change each of these lengths to millimetres.

a) 15 cm   **b)** 1·5 cm   **c)** 4·625 m

a) 15 × 10 = 150 mm   1 cm = 10 mm, so multiply by 10.
                             A whole number, so add a zero.

b) 1·5 × 10 = 15 mm   As in **a)** but this time you simply move
                             the digits 1 place to the left.

c) 4·625 × 1000 = 4625 mm   1 m = 1000 mm so multiply by 1000.

EXAMPLE 4

Change each of these lengths to centimetres.

a) 5·25 m   **b)** 2·542 m   **c)** 20 mm

d) 57 mm   **e)** 42·5 mm

a) 5·25 × 100 = 525 cm   1 m = 100 cm, so multiply by 100,
                               which moves the digits 2 places to the
                               left.

b) 2·542 × 100 = 254·2 cm   The same as in **a)**.

c) 20 ÷ 10 = 2 cm   1 cm = 10 mm, so divide by 10. You
                               need to delete a zero.

d) 57 ÷ 10 = 5·7 cm   The same as in **c)** but this time simply
                               move the digits 1 place to the right.

e) 42·5 ÷ 10 = 4·25 cm   The same as in **d)**.

║ EXAMPLE 5

Change each of these lengths to metres.

a) 148 cm

b) 291·4 cm

c) 3360 mm

a) 148 ÷ 100 = 1·48 m

1 m = 100 cm, so divide by 100, which moves the digits 2 places to the right.

b) 291·4 ÷ 100 = 2·914 m

The same as in **a)**.

c) 3360 ÷ 1000 = 3·36 m

1 m = 1000 mm, so divide by 1000. Move the digits 3 places to the right. The zero at the end can be left off after the decimal point.

║ EXAMPLE 6

Put these lengths in order, smallest first.

3·25 m   415 cm   302 mm   5012 mm   62 cm

To put these in order, first change them all to the same unit. Usually, it is easiest to change to the smallest unit, which is millimetres in this case.

3·25 m = 3·25 × 1000 mm = 3250 mm

Multiply by 1000 to change m to mm.

415 cm = 415 × 10 mm   = 4150 mm

Multiply by 10 to change cm to mm.

302 mm   = 302 mm

5012 mm   = 5012 mm

62 cm = 62 × 10 mm   = 620 mm

So the order is 302 mm, 620 mm, 3250 mm, 4150 mm, 5012 mm

or   302 mm, 62 cm, 3·25 m, 415 cm, 5012 mm.

EXERCISE 8.1

1 Which metric unit would you use to measure each of these?
  a) The width of a book
  b) The height of a room
  c) The width of an envelope
  d) The distance from London to Glasgow
  e) The distance round a running track
  f) The length of a classroom
  g) The length of a finger
  h) The length of a motor car race
  i) The length of a swimming pool
  j) The height of a church steeple
  k) The length of a nail
  l) The width of a window
  m) The distance round your waist
  n) The length of a cross-country race

2 Change each of these lengths to millimetres.
  a) 4 cm   **b)** 33 cm
  c) 2·5 cm   **d)** 52 cm
  e) 4·52 cm   **f)** 2 cm
  g) 4·5 cm   **h)** 9·35 cm
  i) 219 cm   **j)** 99·1 cm

3 Change each of these lengths to millimetres.
  a) 9 m   **b)** 1·129 m
  c) 3·1 m   **d)** 0·3 m
  e) 2·101 m   **f)** 3 m
  g) 2·239 m   **h)** 9·1 m
  i) 4·3 m   **j)** 0·124 m

4 Change each of these lengths to centimetres.
  a) 4 m   **b)** 5·22 m
  c) 9·16 m   **d)** 8·275 m
  e) 52 m   **f)** 5 m
  g) 2·32 m   **h)** 18·16 m
  i) 3·295 m   **j)** 4·1952 m

5 Change each of these lengths to centimetres.
  a) 20 mm   **b)** 140 mm
  c) 35 mm   **d)** 94·6 mm
  e) 660 mm   **f)** 70 mm
  g) 310 mm   **h)** 46 mm
  i) 8000 mm   **j)** 1480 mm

6 Change each of these lengths to metres.
  a) 142 cm   **b)** 4570 cm
  c) 9124 mm   **d)** 5800 mm
  e) 2146·3 mm   **f)** 5142 mm
  g) 570 cm   **h)** 1146 mm
  i) 580·4 cm   **j)** 41 623 mm

7 Write each set of lengths in order of size, smallest first.
  a) 2·42 m
      1600 mm
      284 cm
      9 m
      31 cm
  b) 423 cm
      6100 mm
      804 cm
      3·2 m
      105 mm

Measuring lengths

Look at this ruler.

The numbers on the scale are the centimetre (cm) marks. These are also the longest marks on the scale. The shortest marks on the scale are the millimetre (mm) marks.

The line measures 4·7 cm. That is 4 cm and 7 mm.

> **EXAM TIP**
>
> Notice how the start of the scale is *not* at the end of the ruler. When you measure lines, always make sure that the start of the *scale* is at the end of the line.

EXERCISE 8.2

1 Measure the length of each of these lines in centimetres.

a) ——————————————
b) ————————
c) ——————————————————
d) ——————
e) ————————————————
f) ——————————————————
g) ————————
h) ——
i) ————————————
j) ——————————————————

2 Measure the length of each of these objects in centimetres.

a)

b)

c)

d)

8

3 Measure the length and width of each of these stamps in centimetres.

a)

b)

c)

A **ACTIVITY 2**

Work in pairs.

■ Measure the length of your middle finger. Do it as accurately as you can.

■ Now get your partner to measure your middle finger.

Did you both get the same answers? If not, discuss why they are different.

A **ACTIVITY 3**

Work in pairs.

■ Discuss with your partner how you can measure the length of your foot.

■ Use your method to measure your own foot and your partner's.

■ Check each other's measurements.

STAGE
1

Perimeter

> The perimeter of a shape is the distance all the way around the edge of the shape.

Since the perimeter of a shape is a length, you must use units such as centimetres (cm), metres (m) or kilometres (km).

Simple perimeters

EXAMPLE 7

Find the perimeter of this shape.

5 cm

2 cm   2 cm

5 cm

Perimeter = 5 + 2 + 5 + 2 = 14 cm

Shapes can be irregular, but you work out the perimeter in exactly the same way.

EXAMPLE 8

Find the perimeter of this shape.

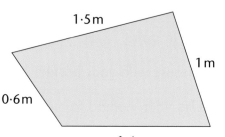

1·5 m

1 m

0·6 m

1·4 m

Perimeter = 0·6 + 1·4 + 1 + 1·5 = 4·5 m

EXAM TIP
Although you don't have to give the units in your working, you must remember to give the units with your answer.

EXERCISE 8.3

Find the perimeter of each of these shapes.

1

3 cm
8 cm   8 cm
3 cm

2
3·5 cm
1·5 cm   1·5 cm
3·5 cm

3
14 cm
5 cm

4
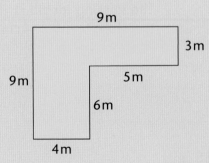
9 m
3 m
9 m   5 m
6 m
4 m

5
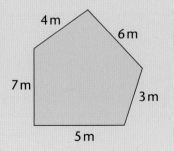
4 m
6 m
7 m
3 m
5 m

6
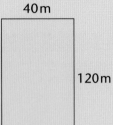
40 m
120 m

7
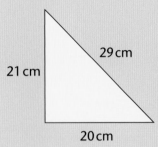
29 cm
21 cm
20 cm

8
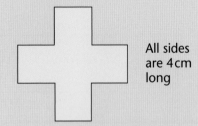
All sides are 4 cm long

9 A square with sides of 25 cm.

10 An equilateral triangle with sides of 14 cm.

> **Hint:**
> An equilateral triangle is one with all its sides the same length.

STAGE
1

More complicated perimeters

Sometimes not all of the lengths of the shape are given in the diagram.

Before trying to find the perimeter, all the lengths should be found.

EXAMPLE 9

Find the perimeter of this shape.

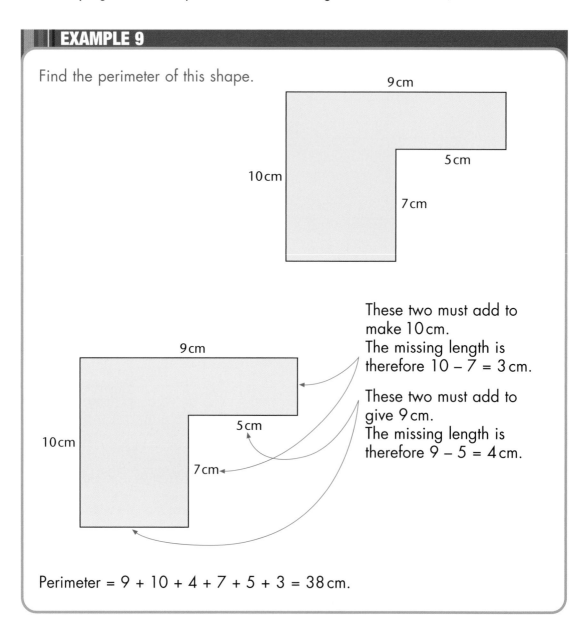

These two must add to make 10 cm.
The missing length is therefore 10 − 7 = 3 cm.

These two must add to give 9 cm.
The missing length is therefore 9 − 5 = 4 cm.

Perimeter = 9 + 10 + 4 + 7 + 5 + 3 = 38 cm.

EXERCISE 8.4

1 Copy each of these diagrams. Find any missing lengths and mark them on your diagram. Find the perimeter of each shape.

a)

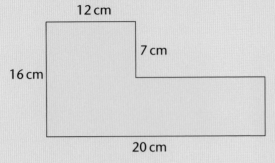

12 cm

7 cm

16 cm

20 cm

b)

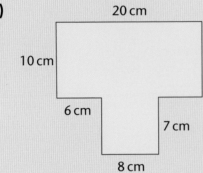

20 cm

10 cm

6 cm

7 cm

8 cm

c)

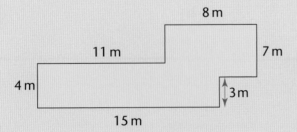

8 m

11 m

7 m

4 m

3 m

15 m

d)

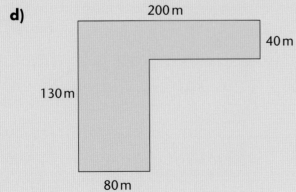

200 m

40 m

130 m

80 m

e)

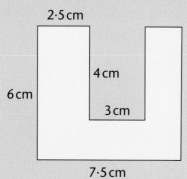

2·5 cm

4 cm

6 cm

3 cm

7·5 cm

f)

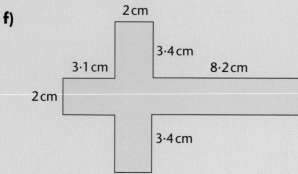

2 cm

3·4 cm

3·1 cm   8·2 cm

2 cm

3·4 cm

2 Measure each of these shapes accurately. Work out the perimeter of each shape.

a)

b)

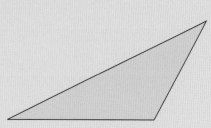

c)

d)

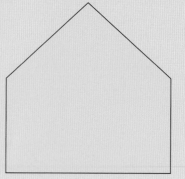

Units, perimeter, area and volume

STAGE
1

EXERCISE 8.4 continued

3 The perimeter of a rectangle is 26 cm.
One side is 8 cm long.
How long is the other side?

> **Hint:**
> Draw a diagram and label as many lengths as possible.

4 A square has a perimeter of 120 cm.
How long is each side?

Area

The area of a two-dimensional shape is the amount of flat space inside the shape.

When you find the area of small shapes you measure the area in square centimetres. This is usually written as cm^2.

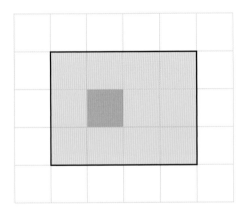

The grid is drawn in centimetres.

The shaded area is a one centimetre square.
Its area is $1\,cm^2$.

In the same way, a square with sides of 1 kilometre will have an area of $1\,km^2$.

STAGE
1

EXAMPLE 10

Find the area of this shape.

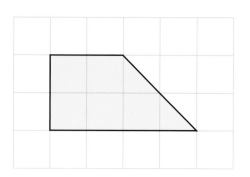

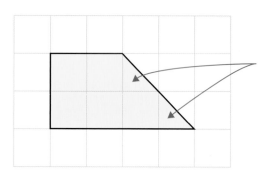

You can fit these two half squares together to make a whole square.

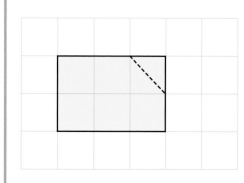

There are 5 + 1 = 6 whole squares altogether.
The area is $6 \, cm^2$.

EXERCISE 8.5

Find the area of each of these shapes by counting squares.
Write your answers in square centimetres.

1 a)

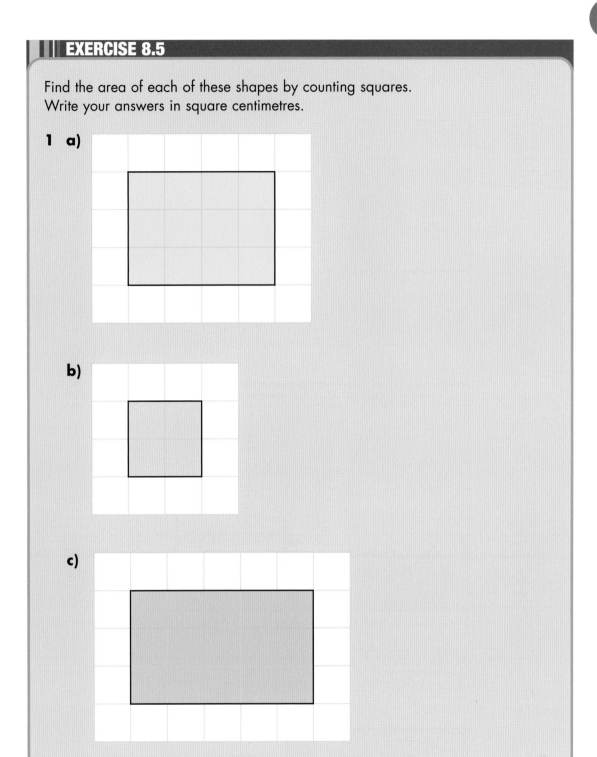

b)

c)

STAGE
1

EXERCISE 8.5 continued

d)

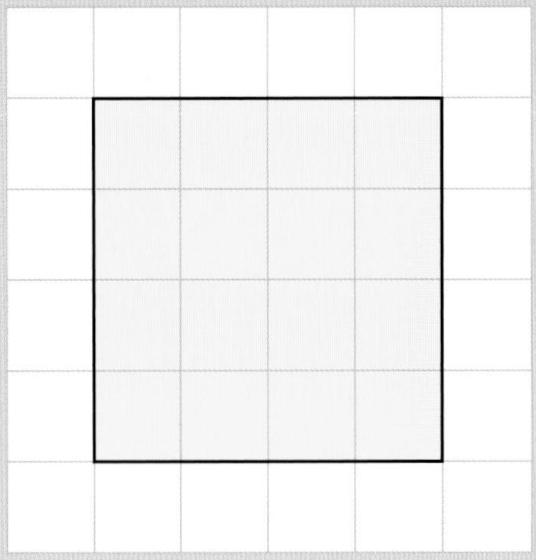

e)

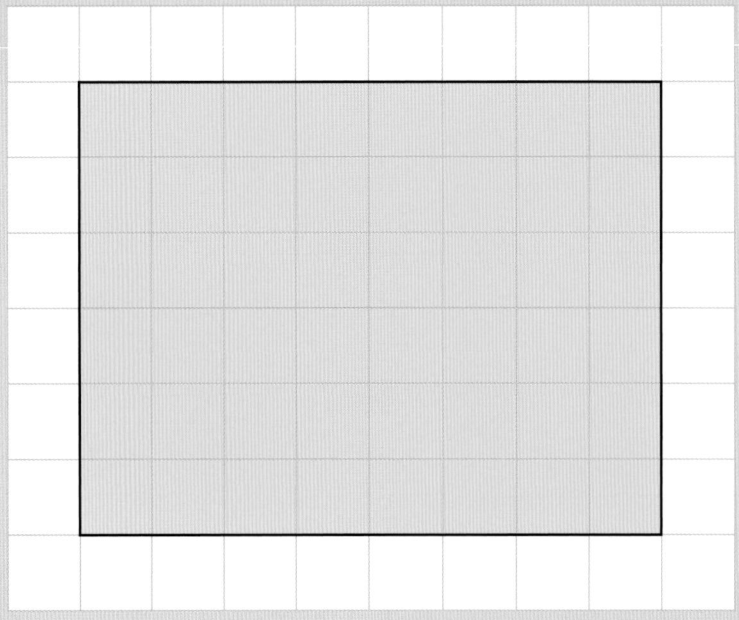

f)

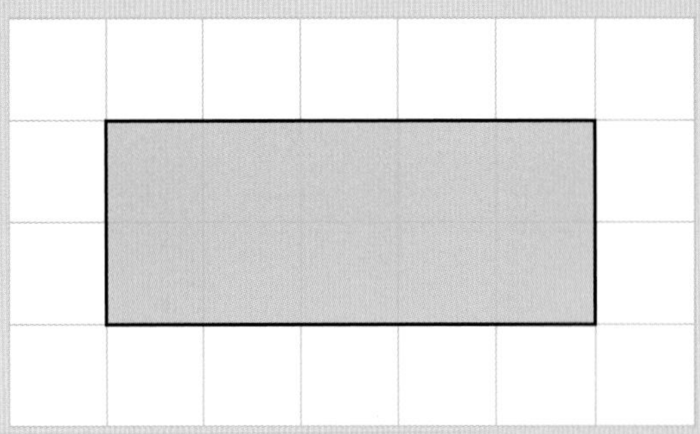

g)

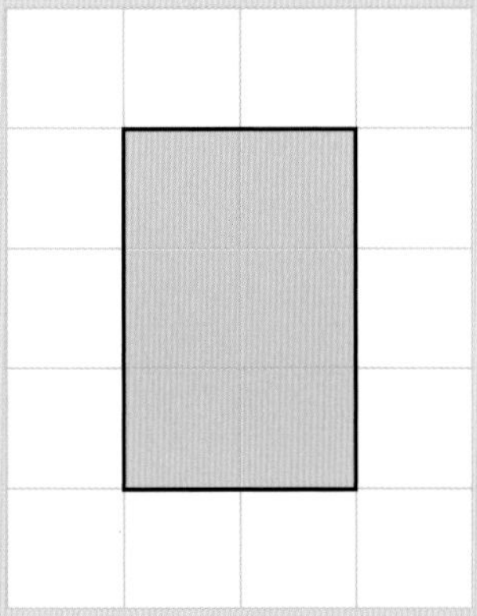

h)

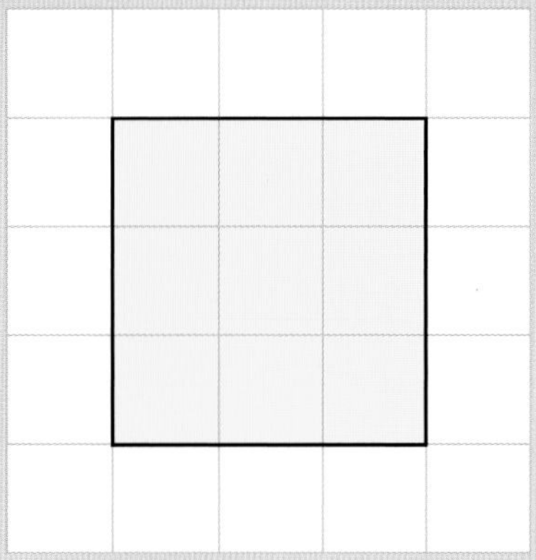

i)

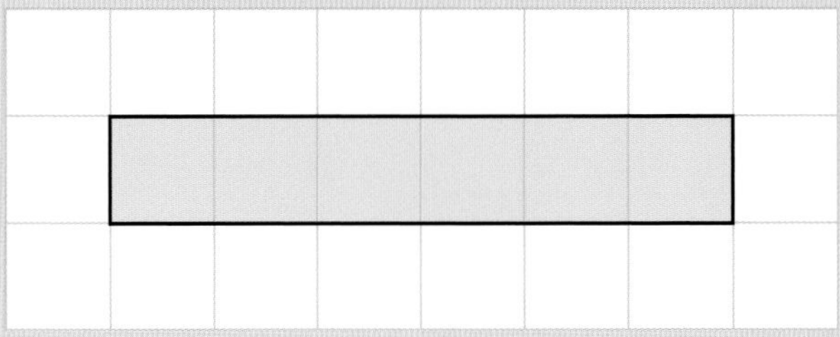

j)

STAGE
1

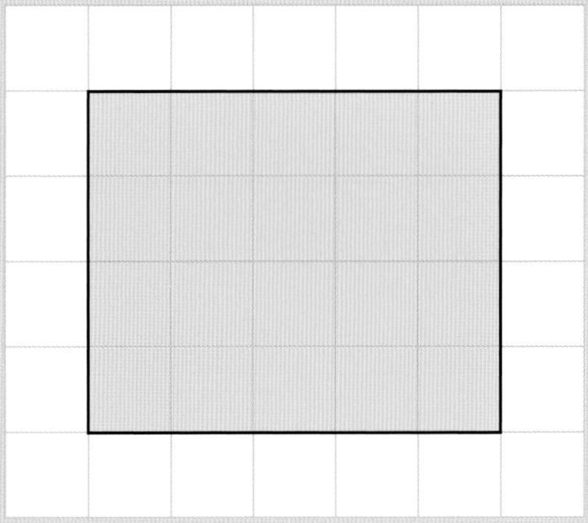

2 a)

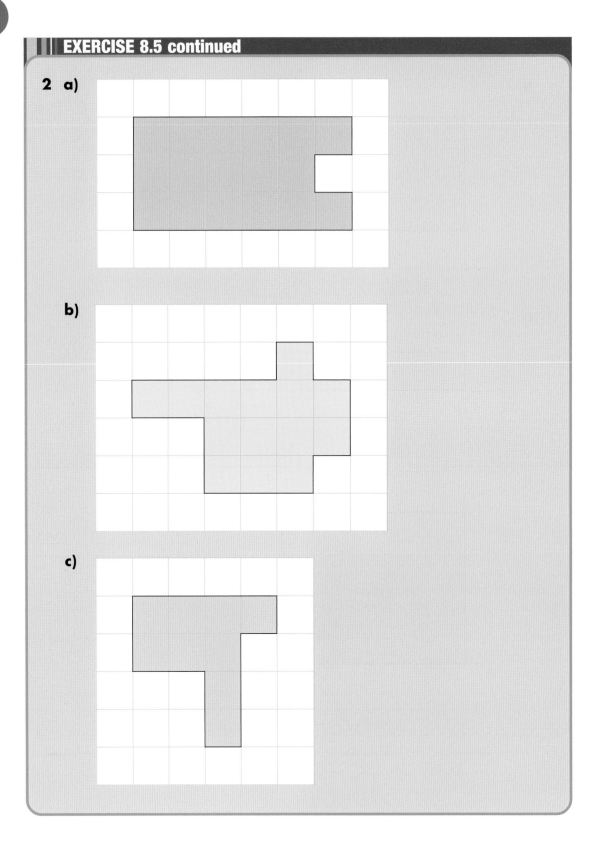

b)

c)

d)

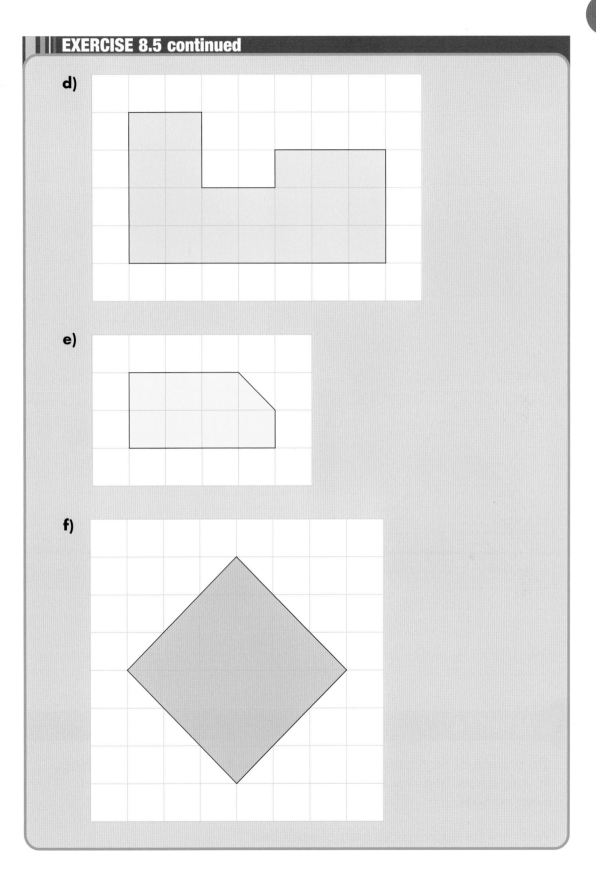

e)

f)

EXERCISE 8.5 continued

g)

h)

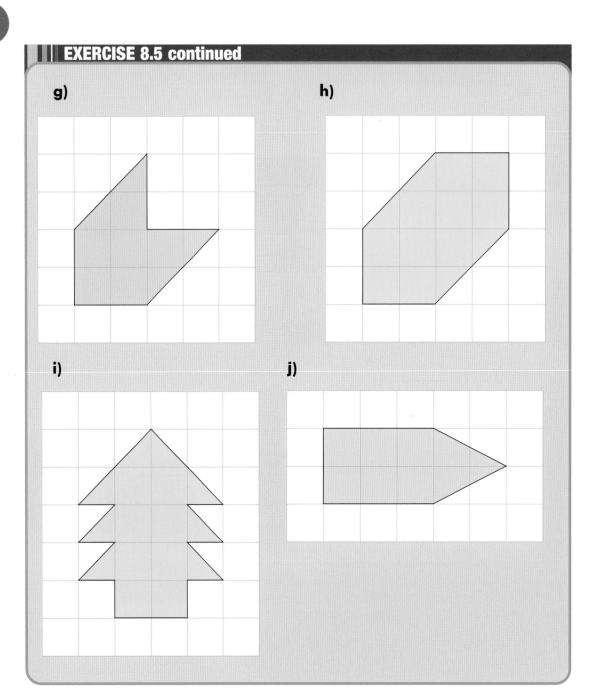

i)

j)

A   ACTIVITY 4

On a piece of centimetre-square paper, draw around some of the things in your school bag. For example, a calculator, a pencil case and an exercise book.

Work out the approximate area of each item.

C CHALLENGE 1

a) How many different shapes can you draw with a perimeter of 18 cm?

b) How many different rectangles can you draw with a perimeter of 20 cm? Which rectangle has the largest area?

Estimating the area of irregular shapes

Often, shapes cover irregular areas. Although it is difficult to find the area exactly, you can find an estimate of the area it covers by counting squares. First count all the full squares. Then make an estimate of the rest of the area by counting bits that are larger than half a square as a full square and ignoring bits that are less than half a square. If some bits are exactly half a square, combine each two to make 1 full square.

▍ EXAMPLE 11

Find the approximate area of this shape.

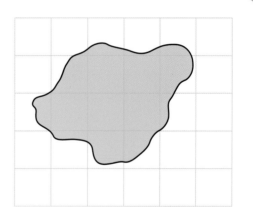

There are 5 full squares (marked F) and 4 larger than half a square (marked L).

A sensible estimate would be 5 + 4 = 9 squares.

So the area is about 9 cm².

> **EXAM TIP**
> When estimating areas, do not try to be too accurate. In Example 11, any answer between 8 and 10 would be acceptable.

STAGE
1

EXERCISE 8.6

Estimate the area of each of these shapes.

1

2

3

4

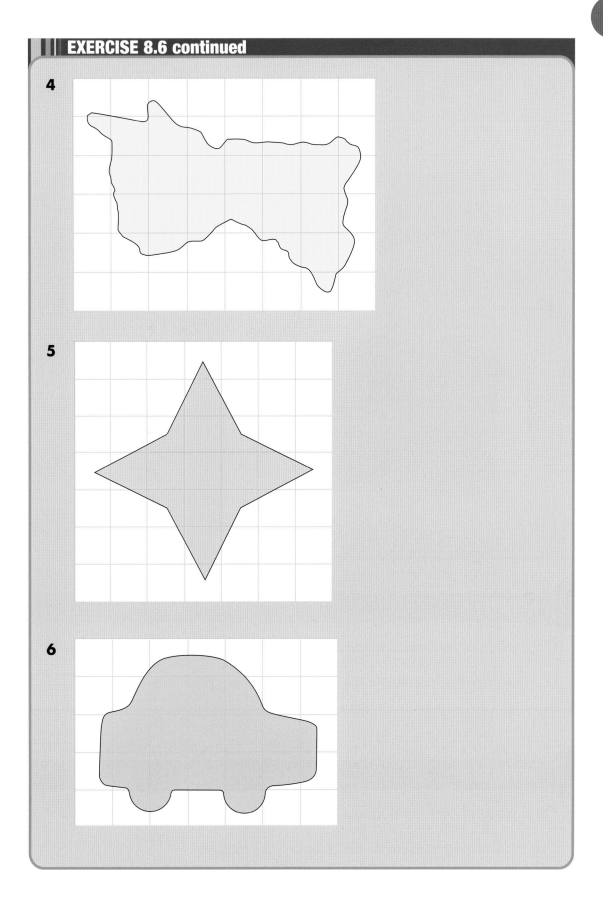

5

6

EXERCISE 8.6 continued

7

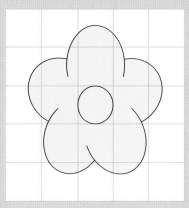

8

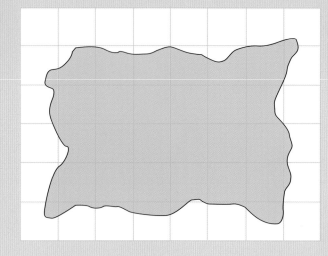

 ACTIVITY 5

On a piece of centimetre-square paper, draw around your hand.

Use this drawing to estimate the area of your handprint.

 ACTIVITY 6

On a piece of centimetre-square paper, draw around your foot.

Use this drawing to estimate the area of your footprint.

Volume

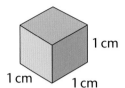

The volume of a three-dimensional shape is the amount of space it takes up.

A cube 1 cm long, 1 cm wide and 1 cm high will have a volume of one cubic centimetre – 1 cm^3.

In the same way, a cube with sides of 1 m will have a volume of one cubic metre – 1 m^3.

The capacity of a box is how much volume is contained within it.

EXAMPLE 12

Find the volume of this solid shape.

One layer has 6 cubes. There are 4 layers.

The volume is $6 \times 4 = 24 \, \text{cm}^3$.

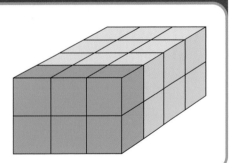

EXAMPLE 13

Find the volume of this solid shape.

One layer has 4 cubes. There are 3 layers.

The volume is $4 \times 3 = 12 \, \text{cm}^3$.

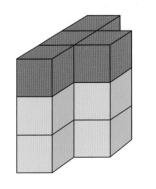

EXERCISE 8.7

1 In these diagrams, each small cube is 1 cm long, 1 cm wide and 1 cm high. Find the volume of each of the solid shapes.

a)

b)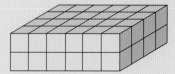

c)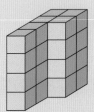

d)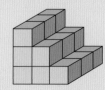

e)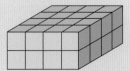

f)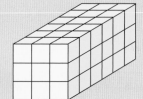

g)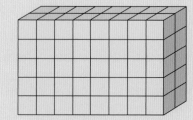

h)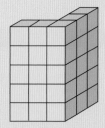

i)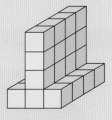

j)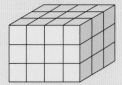

2 How many cubes of volume 1 m³ can fit into a rectangular space measuring 2 m by 3 m by 5 m?

3 How many cubes of volume 1 m³ can fit into a rectangular space measuring 4 m by 4 m by 4 m?

KEY IDEAS

- Lengths can be measured in kilometres (km), metres (m), centimetres (cm) and millimetres (mm).

- There are 1000 metres in a kilometre.

- There are 100 centimetres in a metre.

- There are 10 millimetres in a centimetre.

- The perimeter is the distance all the way around the edge of a shape.

- Perimeters can be measured in kilometres (km), metres (m), centimetres (cm) and millimetres (mm).

- The area of a two-dimensional shape is the amount of flat space inside the shape.

- You can find an estimate of the area inside an irregular shape by counting squares.

- Areas can be measured in square kilometres (km^2), square metres (m^2), square centimetres (cm^2) and square millimetres (mm^2).

- The volume of a three-dimensional shape is the amount of space it occupies.

- Volumes can be measured in cubic metres (m^3), cubic centimetres (cm^3) and cubic millimetres (mm^3).

- The capacity of a box is how much volume is contained within it.

Revision exercise D1

1 Copy and complete these sentences.
  a) The distance from the centre of a circle to its edge is the ....
  b) The .... is the distance round a circle.
  c) A .... is a line across a circle, passing though the centre.

2 Draw a circle of radius 5 cm.
Draw a diameter of the circle.
What is its length?

3 This four-pointed star has eight equal sides.
Explain why it is not a regular octagon.

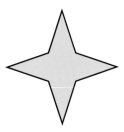

4 What is the special name for a regular quadrilateral?

5 Construct a regular hexagon (6 sides).

6 a) Make a three times enlargement of this shape on centimetre-square paper.

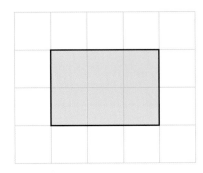

b) Make a two times enlargement of this shape on centimetre-square paper.

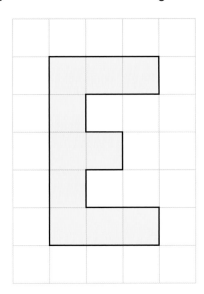

7 Is Shape B an enlargement of Shape A?

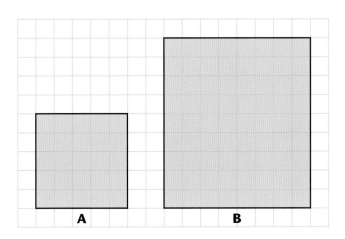

8 What metric units would you use to measure these?
  a) The length of a pencil
  b) The height of a church

9 Change each of these lengths to the units indicated.
  a) 3·2 m to millimetres
  b) 4·5 m to centimetres
  c) 15 cm to millimetres
  d) 584·2 cm to metres
  e) 14 523 mm to metres

STAG
1

10 Put these lengths in order of size, smallest first.

5 m   500 mm   655 cm   7100 mm   2·3 m

11 Find the perimeter of each of these shapes.

a)

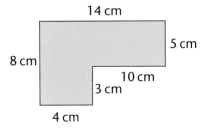

14 cm

5 cm

8 cm

10 cm

3 cm

4 cm

b)

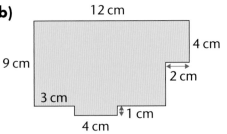

12 cm

4 cm

9 cm

2 cm

3 cm

1 cm

4 cm

c)

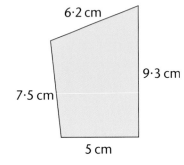

6·2 cm

9·3 cm

7·5 cm

5 cm

12 A piece of A4 paper measures 21 cm by 29·7 cm.
Calculate the perimeter of the piece of paper.

13 Find the area of each of these shapes.

a)

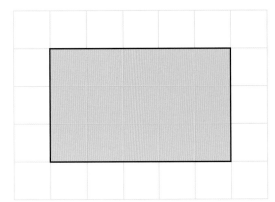

b)

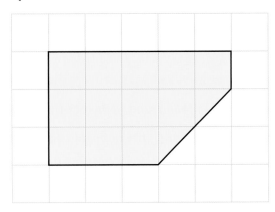

14 Find an estimate of the area of this shape.

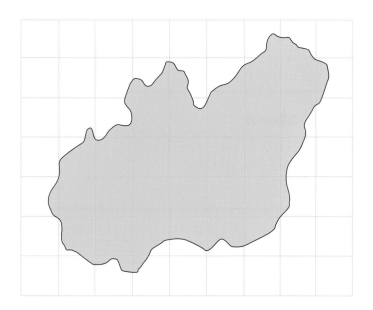

15 In these diagrams, each of the small cubes has a volume of 1 cm³.
Find the volume of each solid shape.

a)

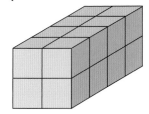

b)

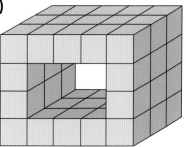

9 Representing data

You will learn about
- Pictogram
- Bar charts and vertical line graphs
- Reading graphs

You should already know
- How to read scales
- How to plot and read points on a graph

Pictograms

Pictograms are graphs that use a symbol to represent a group of units.

The symbols are all the same size and are placed in rows and columns.

EXAMPLE 1

This is an example of a pictogram showing people's favourite colour.

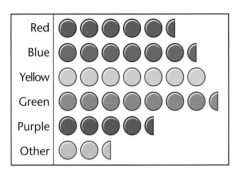

This symbol represents 2 people. ⬤

a) What does ◖ represent?

b) How many people chose blue?

c) How many people altogether were asked about their favourite colour?

a) ⬤ represents 2 people.

◖ is half a symbol so it represents 1 person.

b) There are $6\frac{1}{2}$ symbols for blue.

$6 \times 2 = 12$
$12 + 1 = 13$

13 people chose blue.

c) You need to work out the number for each colour and then add the numbers together.

Red: $5\frac{1}{2}$ symbols so $5 \times 2 + 1$ = 11 people
Blue: $6\frac{1}{2}$ symbols so $6 \times 2 + 1$ = 13 people
Yellow: 7 symbols so 7×2 = 14 people
Green: $7\frac{1}{2}$ symbols so $7 \times 2 + 1$ = 15 people
Purple: $4\frac{1}{2}$ symbols so $4 \times 2 + 1$ = 9 people
Other: $2\frac{1}{2}$ symbols so $2 \times 2 + 1$ = 5 people
Total: $11 + 13 + 14 + 15 + 9 + 5 = 67$ people

Alternatively, you can count all the symbols and multiply by 2.

$33\frac{1}{2}$ symbols so $33 \times 2 + 1 = 67$ people

STAGE
1

Bar charts and vertical line graphs

Another way to illustrate data is to use a **bar chart** or **vertical line graph**.

The next example uses a bar chart.

EXAMPLE 2

The bar chart shows the number of children in each of the 30 families of Class 7S.

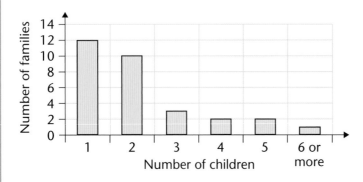

a) How many families have three children?

b) How many families have fewer than three children?

a) The column for three children is 3 high, so three families have three children.

b) Fewer than 3 is 1 or 2.
There are 12 families that have one child.
There are 10 families that have two children.
So there is a total of 22 families with fewer than three children.

Vertical line graphs are sometimes used instead of bar charts.

They are read in the same way.

This vertical line graph shows the same data as in Example 2.

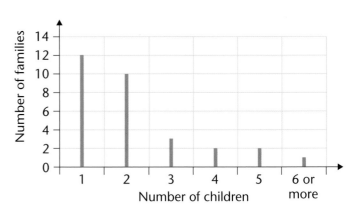

1 Emma drew this pictogram to show the number of books borrowed from the school library one week.

represents 10 books.

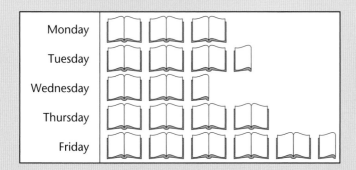

a) How many students borrowed books on each of the days?
b) Which day was the most popular? Why do you think that might be?

2 The pictogram shows how many people in a small village watch each of the different TV channels.

represents 40 people.

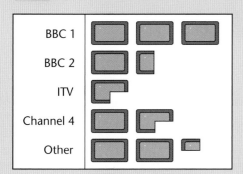

a) Which channel is the most popular?
b) Which channel is the least popular?
c) Work out how many people watch each of the channels.

STAGE
1

3 The bar chart shows the eye colour of students in Class 7F.

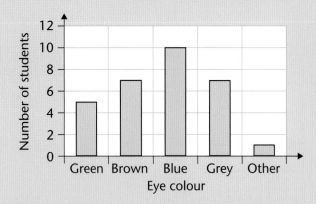

a) How many students have brown eyes?

b) How many more students have blue eyes than have grey eyes?

c) How many students are there in the class?

4 Geta is doing a survey on the number of letters in a person's first name. The table shows the data she collected.

Number of letters	Number of students
4	6
5	5
6	9
7	2
8	4
9	1

a) Draw a pictogram to show this information.

Use the symbol ☺ to represent 2 students.

b) Which is the most common length of first name?

EXERCISE 9.1 continued

5 The pictogram shows the number of CDs sold in a music shop in the week before Christmas.

represents 20 CDs.

Date		Total
21st Dec		
22nd Dec		
23rd Dec		
24th Dec		

a) What numbers should go in the 'Total' column?

b) What were the total sales?

6 The table shows the numbers of students absent from school one week.

Day	Total
Monday	8
Tuesday	5
Wednesday	7
Thursday	10
Friday	6

Draw a bar chart to represent this information.

7 The table shows the numbers of bikes sold by 'Bikes 'R' Us' in four weeks.

Week	Total
Week 1	12
Week 2	10
Week 3	19
Week 4	13

Draw a pictogram to represent this information.

Use the symbol to represent 8 bikes.

STAGE
1

A ACTIVITY 1

Conduct a survey of the students in your class.

Ask everyone something like

- 'Are you left- or right-handed?'
- 'What month were you born in?'
- 'How many letters are there in your first name?'
- 'How many people are in your family?'
- 'What sorts of pets do you have in your family?'
- A question of your own.

Draw a pictogram, bar chart or vertical line graph to show your data.

Reading from graphs

Emily is in hospital. Every 3 hours her temperature is taken.

Time	00:00	03:00	06:00	09:00	12:00	15:00	18:00
Temperature (°C)	38	38·5	38·2	39·2	39·2	38·6	39·4

The points are plotted on a graph.

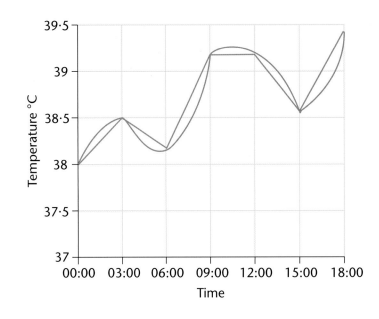

The points are joined by straight lines – but does this mean anything?

Would anyone be able to state that, at 10:00, say, Emily's temperature was 39·2°C? Emily's temperature might really vary like the blue line on the graph.

The straight lines are drawn to give an impression. An intermediate value can only give you an estimate of her temperature at a time between the times when it was measured; it is not accurate.

EXERCISE 9.2

1 This graph shows the average monthly rainfall for a town in Britain measured over a 30-year period.

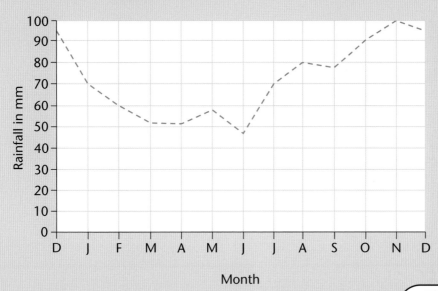

Month

> **EXAM TIP**
> The points are joined with a dashed line, as between the months has no meaning.

a) What was the lowest rainfall?
b) What was the difference between the highest and the lowest rainfall?
c) Which two months had an average rainfall of 70 mm?
d) Are you able to say with certainty what the rainfall is in the middle of each month?

STAGE
1

EXERCISE 9.2 continued

2 Julie carried out a survey of how many people were in each car passing the school gates between 08:30 and 09:00.
Here is the graph of her results.

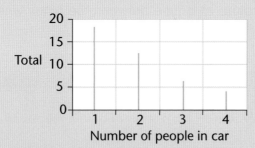

Why is it not sensible to join the tops of the vertical lines?

3 Peggy is the caretaker at a school.
She makes sure the heating in the school is kept at a suitable temperature.
She takes the temperature several times each day and draws graphs like this.

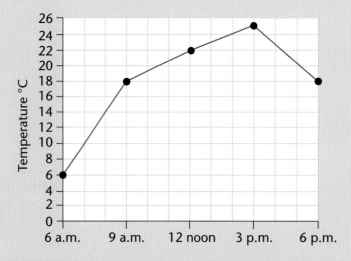

a) What was the highest temperature on the day she drew this graph?
b) What was the temperature at 9 a.m.?
c) Roughly, what was the temperature at 5 p.m.?

4 Eddie makes a cup of tea and measures the temperature as the tea cools down.

His measurements are given in the table.

Time (seconds)	0	20	40	60	80	100	120
Temperature (°C)	100	92	76	46	28	16	10

a) Plot these points on a graph and join them with straight lines.

b) Use your graph to find

    (i) roughly what the temperature was after 50 seconds.

    (ii) roughly how long it took the tea to cool down to 85°C.

5 The sales in a local shop over the course of a week are given in the table.

Day	Monday	Tuesday	Wednesday	Thursday	Friday	Saturday
Sales (£)	535	418	502	289	612	876

a) Plot these points on a graph and join them with dashed straight lines.

b) Is it sensible to read between Monday and Tuesday?

c) If the shop wanted to open on Sunday, could you use the graph to predict what their sales might be?

6 The amount of a chemical produced in an industrial process is shown below.

Time	08:00	10:00	12:00	14:00	16:00	18:00
Amount (kg)	0	24	52	91	112	124

a) Plot these points on a graph and join them with straight lines.

b) About how much would be produced

    (i) by 11:30?

    (ii) by 17:00?

c) When is the greatest amount produced in a two-hour period?

d) By about what time would 80 kg be produced?

STAGE 1

CHALLENGE 1

Look in newspapers or magazines and find some examples of pictograms, bar chart and other graphs representing data.

Make a wall display explaining what they show.

KEY IDEAS

- Symbols in a pictogram are the same size and are placed in rows and columns.

- The key to a pictogram tells you 'how many' each symbol represents.

- Part of a symbol can be used to represent a fraction of the number.

- You can use a bar chart or a vertical line graph to represent data. The height of the bar or line tells you 'how many' of the category there are.

- You can use a line graph to represent data.

- The lines between points on a graph are drawn to show the trend. Readings from the lines are estimates only. They cannot be read accurately.

- Sometimes there is no meaning between the plotted points, then the line is dashed.

Listing 10

You will learn about

- Making lists of arrangements

You should already know

- That order can be important

Three friends, Anne, Bill and Chloe, go to the cinema. The film is so popular that there are just three seats left, all next to each other. The friends can't decide who will sit where.

EXAM TIP

Use letters for the names. This will make listing easier.

Write down the different ways that they can sit next to each other.

Use a systematic way of listing.

For example, keep A in seat 1 and change the others.
Then keep B in seat 1, etc.

There are six possible ways they can sit.

```
A B C
A C B
B A C
B C A
C A B
C B A
```

STAGE
1

You could also use a table to list the different ways.

Seat 1	Seat 2	Seat 3
A	B	C
A	C	B
B	A	C
B	C	A
C	A	B
C	B	A

EXERCISE 10.1

1 How many two-digit numbers can you make using the digits 5 and 7?

2 I have a box of red counters and a box of blue counters.
How many different two-counter patterns can I make?

3 Dave, Emma and Fred sit in three seats next to each other in assembly. Emma won't sit in the middle.
List the different ways they can sit.

4 How many different three-digit numbers can you make using the digits 1, 2 and 3?

5 I have three jumpers. One is yellow, one is green and one is black.
I wear a different jumper each day for three days.
How many different orders are there in which to wear the jumpers?

6 A school has the same three meals on offer each day: Pizza, Burger and Salad.
List all the possible selections of meal you can have in three days, for example Pizza, Salad, Pizza.

7 List the different ways of arranging the letters P, Q, R and S using each letter once only.

Hint:
There are 24 different ways.

8 Four people, A, B, C and D, work in an office.
They must take their annual holiday one at a time.
D wants to be last and A and B want to follow on after each other.
List the possible orders of the holidays.

9 I have a box of green and red counters.
How many different patterns can I make using three counters?

10 How many four-counter patterns can you make with green and red counters only?

A **ACTIVITY 1**

List all the three-digit numbers can you make using the digits 4, 5 and 6

a) if you can only use each digit once.

b) if you can use each digit as many times as you like.

C **CHALLENGE 1**

I have four shirts. Two are white and two are blue.

I wear a different shirt each day for four days.

List the different *colour* orders in which I can wear the shirts.

C **CHALLENGE 2**

Elaine has two skirts. One is blue and one is red.

She has three blouses: one grey, one yellow and one white.

How many different outfits can she make?

STAGE
1

C CHALLENGE 3

Mum gives us a choice for Sunday lunch.

Meat	Beef, Pork, Chicken
Potatoes	Roast, Boiled
Vegetables	Peas, Cabbage

How many different meals can we choose?

We can only choose one item from each list.

C CHALLENGE 4

How many four-digit numbers can you make using 3, 4, 5 and 6

a) if you can only use each digit once?

b) if you can use each digit as many times as you like?

Can you find a way to do this *without* writing them all down?

K KEY IDEAS

- When making lists it is easier to use letters or some other shorthand to represent the items.

- A systematic approach is best to ensure that all possible arrangements have been found.

Revision exercise E1

1 The pictogram below represents the income of an ice-cream seller during one week in July.

 represents £10.

Day of the week		Income (£)
Monday	♦♦♦♦♦	
Tuesday	♦♦♦◗	
Wednesday		60
Thursday	♦◗	
Friday		45
Saturday	♦♦♦♦♦♦♦	
Sunday		85

a) Copy and complete the pictogram.
b) On which day did he sell the most ice-creams?
c) How much more money did he take on Wednesday than on Tuesday?
d) How much money did he take altogether?
e) Give one possible reason why he sold so little ice-cream on Thursday and one possible reason why he sold so much on Saturday.

E1

2 This bar chart shows the results for a local football team for one season.

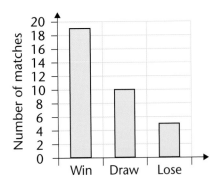

a) How many matches did the team play altogether?
b) The table below shows the results for a second team.
Draw a bar chart to show the results for this team.

Result	Number of matches
Win	8
Draw	7
Lose	3

3 The table below shows the width of a tree as it grows.

Age (years)	10	20	30	40	50
Width (cm)	10	16	25	41	63

a) Plot these points on a grid and join them with straight lines.
b) Use your graph to find
   (i) the width of the tree after 35 years.
   (ii) how old the tree was when its width was 20 cm and when its width was 50 cm.

4 Copy and complete this table to show the possible ways the coins can land when you toss a 5p, a 10p and a £1 coin together. You will need more rows.

5p	10p	£1
H	H	H

5 List the different ways of arranging the letters A, B, C and D using each letter once only.

Stage 2 Contents

STAGE
2

Contents

STAGE
2

Using decimals

You will learn about

- Place value in decimals
- How to add and subtract measurements and money

You should already know

- How to change between units of length
- The meaning of place value in integers
- How to order integers by size

Place value in decimals

Look at a ruler.

Some rulers are marked in millimetres, like this.

The arrow points to 38 mm.

This is between 30 mm and 40 mm and is $\frac{8}{10}$ of the way from 30 mm towards 40 mm.

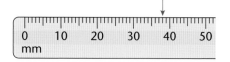

Some rulers are marked in centimetres, like this.

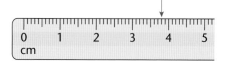

The arrow points to 3·8 cm.

This is between 3 cm and 4 cm and is $\frac{8}{10}$ of the way from 3 cm towards 4 cm.

ACTIVITY 1

■ Find the position of 3·8 cm or 38 mm on your ruler.

■ Then find the position of 7·4 cm or 74 mm on your ruler.
Work in pairs and describe its position in a similar way to page 151.

■ Choose some more measurements that are on your ruler.
Describe their positions on the ruler in a similar way.

Decimals are a way of describing numbers which are not integers.

We use them when measuring lengths, as in Activity 1.

We also use them in connection with money. For instance, £0·42 means 42p, or $\frac{42}{100}$ of a pound.

ACTIVITY 2

■ Think of other situations where you know that decimals are used.

■ How many can your group find?

This table shows place values, including decimals.

Th	H	T	U	.	$\frac{1}{10}$	$\frac{1}{100}$	$\frac{1}{1000}$
		7	4				
			0	.	4	2	
			3	.	8		

It shows some of the numbers you have met in this chapter so far.

It can be used with other numbers you meet.

The table tells you that 0·42 means 0 units, $\frac{4}{10}$ and $\frac{2}{100}$.

But $\frac{4}{10}$ is the same as $\frac{40}{100}$. So 0·42 can just be written as $\frac{42}{100}$.

The table also tells you that 3·8 means 3 units and $\frac{8}{10}$.

This is written as $3\frac{8}{10}$.

A mixture of a whole number and a fraction is called a **mixed number**.

EXAMPLE 1

What is the place value of the digit 4 in each of these numbers?

a) 74 000
c) 8·415

b) 643·2
d) 0·04

Use the place value table.

Ten Th	Th	H	T	U	.	$\frac{1}{10}$	$\frac{1}{100}$	$\frac{1}{1000}$
7	(4)	0	0	0				
		6	(4)	3	.	2		
				8	.	(4)	1	5
				0	.	0	(4)	

a) 4 thousands
c) 4 tenths

b) 4 tens
d) 4 hundredths

EXERCISE 1.1

1 Write in words the place value of the digit 4 in each of these numbers.
   a) 40   **b)** 0·4
   c) 40 000   **d)** 8·74
   e) 0·014

2 What is the place value of the digit 6 in each of these numbers?
   a) 6000   **b)** 4·6
   c) 8462   **d)** 9·46

3 Write each of these decimals as a fraction or a mixed number.
   a) 0·6   **b)** 4·3
   c) 14·1   **d)** 0·75
   e) 9·03

STAGE
2

4 Write each of these amounts of
money in pounds.
a) 142p   **b)** 92p
c) 607p   **d)** 5p
e) 7p   **f)** 860p

5 Write each of these amounts of
money in pence.
a) £0·87   **b)** £1·56
c) £0·08   **d)** £0·26

6 Write each of these lengths in metres.
a) 12 cm   **b)** 874 mm
c) 21·8 cm   **d)** 56 mm
e) 138 cm

7 Write each of these lengths in
centimetres.
a) 2·36 m   **b)** 83 mm
c) 0·57 m   **d)** 5·8 m
e) 470 mm

Adding and subtracting decimals

A   ACTIVITY 3

■ Use your ruler to draw a line which is 8·5 cm long.
  Mark on the line a point which is 4·7 cm from one end.
  Measure the distance from this point to the other end of your line.

■ Do an appropriate addition or subtraction to find out if your measurement
  is accurate.

■ Work in pairs.
  Repeat the instructions above using different measurements.
  One person finds the distance by drawing, the other by doing a
  subtraction.

For a line 75 mm long and a point 52 mm from one end, the distance from the
point to the other end is found by doing this subtraction.

```
  75
– 52
─────
  23 mm
─────
```

Working in centimetres, for a line 7·5 cm long and a point 5·2 cm from one end, the distance from the point to the other end is found by doing this subtraction.

$$
\begin{array}{r}
7·5 \\
- 5·2 \\
\hline
2·3 \text{ cm}
\end{array}
$$

When you use the column method of adding or subtracting, make sure you line up all the decimal points under each other.

Then add or subtract as you would with integers.

EXAMPLE 2

Julie buys a kilogram of apples costing £1·21 and a punnet of strawberries costing 99p.

How much has she spent?

First, make the units the same.   99p = £0·99

Then do the addition.

$$
\begin{array}{r}
1·21 \\
+ 0·99 \\
\hline
2·20
\end{array}
$$

Answer: £2·20

You could also solve this problem by working in pence and then changing your answer into pounds.

EXAM TIP

When working with money you should write the answer with two decimal places, as £2·20 not £2·2.

STAGE
2

EXAMPLE 3

A piece of wood is 2·3 m long. 75 cm is cut off.

How much remains?

First, make the units the same.   75 cm = 0·75 m

Then do the subtraction.

$$\begin{array}{r} 2\cdot30 \\ -\ 0\cdot75 \\ \hline 1\cdot55 \end{array}$$

EXAM TIP

Units of measurements should always be the same.

Answer: 1·55 m

You could also solve this problem by working in centimetres and then changing your answer into metres.

EXERCISE 1.2

1 Work out these.

a) $\begin{array}{r} 6\cdot72 \\ +\ 7\cdot19 \\ \hline \end{array}$

b) $\begin{array}{r} 18\cdot95 \\ +\ 23\cdot14 \\ \hline \end{array}$

c) $\begin{array}{r} 27\cdot54 \\ +\ 83\cdot61 \\ \hline \end{array}$

d) $\begin{array}{r} 5\cdot91 \\ +\ 8\cdot72 \\ \hline \end{array}$

e) $\begin{array}{r} 16\cdot74 \\ +\ 43.97 \\ \hline \end{array}$

f) $\begin{array}{r} 33\cdot51 \\ +\ 79\cdot86 \\ \hline \end{array}$

g) $\begin{array}{r} 6\cdot82 \\ +\ 2\cdot49 \\ \hline \end{array}$

h) $\begin{array}{r} 26\cdot92 \\ +\ 18\cdot54 \\ \hline \end{array}$

i) $\begin{array}{r} 27\cdot36 \\ +\ 91\cdot48 \\ \hline \end{array}$

j) $\begin{array}{r} 9\cdot16 \\ +\ 7\cdot72 \\ \hline \end{array}$

k) $\begin{array}{r} 13\cdot84 \\ +\ 37\cdot67 \\ \hline \end{array}$

l) $\begin{array}{r} 38\cdot53 \\ +\ 89\cdot76 \\ \hline \end{array}$

2 Work out these.

a) 16·78
    – 7·13

b) 28·75
    – 13·84

c) 128·36
    – 73·52

d) 13·49
    – 5·18

e) 47·51
    – 26·74

f) 439·87
    – 218·03

g) 21·74
    – 8·13

h) 36·86
    – 12·78

i) 130·46
    – 83·92

j) 12·59
    – 7·16

k) 35·57
    – 28·74

l) 409·15
    – 213·08

3 Work out these.
  a) £6·84 + 37p + £9·41
  b) £16·83 + 94p + £6·81 + 32p
  c) £61·84 + 76p + £9·72 + £41·32 + 83p
  d) £3·89 + 73p + 68p + £91·80

4 Work out these.
  Give your answers in the larger unit.
  a) 6·1 m + 92 cm + 9·3 m
  b) 3·2 m + 28 cm + 6·74 m + 93 cm
  c) 7·2 m – 165 cm
  d) 8·5 m – 62 cm
  e) 7·6 cm – 8 mm

5 Two pieces of wood are put end to end.
  Their lengths are 2·5 m and 60 cm.
  Find the total length of the wood, in metres.

STAGE
2

6 In the high jump, Angela jumps 1·62 m and Sarah jumps 1·47 m.
Find the difference between the heights of their jumps.

7 The times for the first and last places in a 100-metre race were 11·73 seconds and 14·38 seconds.
Find the difference between these times.

8 Sam buys a magazine for £2·25, a newspaper for 70p and a pack of sweets for 58p.
How much does he spend altogether?

9 Mary has £7·19 in her purse
She spends 70p on parking.
How much does she have left?

10 Winston is buying his lunch.
He buys a sandwich for £1·89, a packet of crisps for 37p, an apple for 40p and a drink for 45p.
He pays with a £5 note.
How much change does he receive?

K KEY IDEAS

- Write amounts of money with two decimal places, e.g. as £4·20 not £4·2.

- You can only add or subtract measurements if they are in the same units.

- To add and subtract decimals, line up the decimal points and then add or subtract as you would with whole numbers.

Number patterns

You will learn about

● Recognising and describing patterns in numbers

You should already know

● How to continue simple number patterns
● What odd and even numbers are

Sequences

You met number patterns in Stage 1.

Here is one number pattern.

6   11   16   21   26   31   ...

You can see that, to get from one number to the next, you add 5.

Here is another number pattern.

1   2   4   8   16   32   ...

To get from one number to the next this time you multiply by 2.

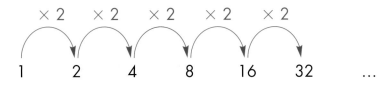

Number patterns like these are called **sequences**.

EXAMPLE 1

Explain how you find the next number in each of these sequences and give the next number.

a) 1   4   7   10   13   16   ...

b) 63   56   49   42   35   28   ...

c) 1 000 000   100 000   10 000   1000   100   ...

a) To get from one number to the next, you add 3.

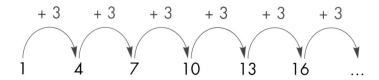

The next number is 16 + 3 = 19.

b) To get from one number to the next, you subtract 7.

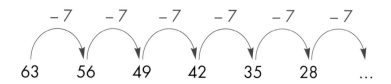

The next number is 28 − 7 = 21.

c) To get from one number to the next, you divide by 10.

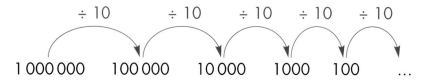

The next number is 100 ÷ 10 = 10.

In the sequences in the next example, some of the numbers are missing.

It is possible to work out what they should be by looking at the numbers in the sequence and working out how to get from one number to the next.

EXAMPLE 2

Find the missing numbers in each of these sequences.

a) 10   17   ☐   ☐   38   45

b) 35   ☐   23   17   ☐   5

c) ☐   15   ☐   23   27   31

a) To get from one number to the next you add 7.

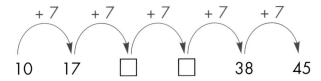

So the first missing number is 17 + 7 = 24.
The second missing number is 24 + 7 = 31.

b) To get from one number to the next you subtract 6.

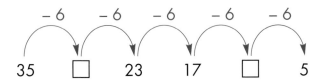

So the first missing number is 35 – 6 = 29.
The second missing number is 17 – 6 = 11.

c) To get from one number to the next you add 4.

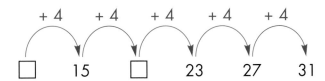

To get to the *previous* number you must subtract 4.
So the first missing number is 15 – 4 = 11.

You find the second missing number by adding 4 in the usual way.
The second missing number is 15 + 4 = 19.

STAGE
2

EXERCISE 2.1

1 Explain how you find the next number in each of these sequences and give the next number.

a) 2   6   10   14   18   22   ...
b) 3   11   19   27   35   43   ...
c) 4   9   14   19   24   29   ...

2 Explain how you find the next number in each of these sequences and give the next number.

a) 33   29   25   21   17   13   ...
b) 23   20   17   14   11   8   ...
c) 102   89   76   63   50   37   ...

3 Explain how you find the next number in each of these sequences and give the next number.

a) 1   3   9   27   81   243
b) 15 625   3125   625   125   25
c) 1   2   4   7   11   16

4 Find the missing numbers in each of these sequences.

a) 7   12   17   ☐   ☐   32   ☐
b) 25   ☐   19   16   ☐   10   ☐
c) 1   4   16   ☐   ☐   1024   4096

5 Explain how you know whether or not 77 is in each of these sequences.

a) 7   14   21   28   35   42
b) 14   18   22   26   30   34

C CHALLENGE 1

Write down a sequence of your own.

Swap with a partner and explain how to find the next number of your partner's sequence.

CHALLENGE 2

Look at this sequence of numbers.

1   4   9   16   25   36

a) Find the next number in the sequence.

b) Explain how you get each number of the sequence.

c) The numbers in this sequence have a special name.
Find out what it is.

More complicated patterns

Look at this pattern.

$1 \times 8 = 8$
$2 \times 8 = 16$
$3 \times 8 = 24$
$4 \times 8 = 32$
$5 \times 8 = 40$
$6 \times 8 = 48$

You will recognise it as the 8 times table.
Now look at the patterns in the numbers as you go down the calculations.
The first number increases by 1 each time.
The second number stays the same.
The last number increases by 8 each time.

Sometimes the patterns are more complicated but you can still look for patterns
in the numbers, as in this example.

STAGE
2

EXAMPLE 3

Look at this pattern.

$9 \times 12 + 3 = 111$
$9 \times 23 + 4 = 211$
$9 \times 34 + 5 = 311$

Explain how you find the next calculation in the pattern and give the next calculation.

$9 \times 12 + 3 = 111$

$+ 11 \quad + 1 \quad + 100$

$9 \times 23 + 4 = 211$

$+ 11 \quad + 1 \quad + 100$

$9 \times 34 + 5 = 311$

The first number stays the same.

The second number increases by 11 each time.

The third number increases by 1 each time.

The answer to the calculation increases by 100 each time.

So the next calculation is

$9 \times 34 + 5 = 311$

$+ 11 \quad + 1 \quad + 100$

$9 \times 45 + 6 = 411.$

EXERCISE 2.2

1 Look at this pattern.

99 – 11 = 88
88 – 22 = 66
77 – 33 = 44

Explain how you find the next calculation in the pattern and give the next calculation.

2 Look at this pattern.

2 × 22 +   6 =   50
4 × 22 + 12 = 100
6 × 22 + 18 = 150

Explain how you find the next calculation in the pattern and give the next calculation.

3 Look at this pattern.

10 × 10 + 11 = 111
20 × 10 + 22 = 222
30 × 10 + ☐ = 333
☐ × 10 + 44 = 444
50 × 10 + 55 = ☐

Find the missing numbers in the pattern.

KEY IDEAS

■ A sequence is a set of numbers linked by a pattern.

STAGE
2

3 Angles

Describing angles

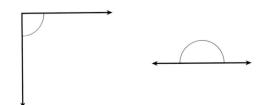

To describe the angle required to rotate one of these arrows on to the other, terms like '$\frac{1}{4}$ turn' and '$\frac{1}{2}$ turn' are accurate enough.

For angles such as these, however, a more accurate measurement is required.

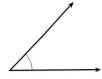 

To do this we use a scale marked in degrees.

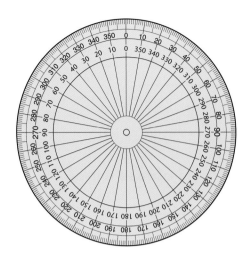

On this scale, one whole turn is equal to 360 degrees. This is written as 360°.

So $\frac{1}{2}$ turn is equal to 180° and $\frac{1}{4}$ turn is equal to 90°.

EXAMPLE 1

What angle in degrees is equal to

a) two whole turns?

b) $\frac{3}{4}$ turn?

a) Two whole turns = 2 × 360
= 720°.

b) $\frac{3}{4}$ turn = $\frac{3}{4}$ × 360°
= 270°.

Angles less than $\frac{1}{4}$ turn (90°) are called **acute** angles.

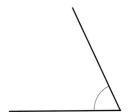

Angles between $\frac{1}{4}$ turn (90°) and $\frac{1}{2}$ turn (180°) are called **obtuse** angles.

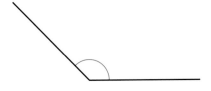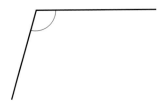

Angles of $\frac{1}{4}$ turn (90°) are called **right** angles.

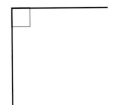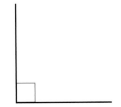

Angles of more than $\frac{1}{2}$ turn (180°) are called **reflex** angles.

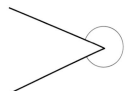

EXERCISE 3.1

1 For each of these angles, say whether it is acute, obtuse, reflex or a right angle.

a)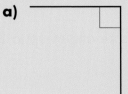

b)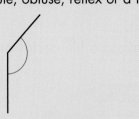

c)

d)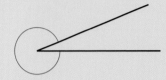

2 What sort of angle is each of these?

a)

b)

c)

d)

e)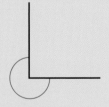

3 Is an angle of each of these sizes acute, obtuse or reflex?
   a) 145°   **b)** 86°
   c) 350°   **d)** 190°

4 What sort of angle has each of these sizes?
   a) 126°   **b)** 226°
   c) 26°   **d)** 90°

3

Measuring angles

The instrument used to measure an angle is called a **protractor** or angle measurer.

Some protractors are full circles and can be used to measure angles up to 360°.

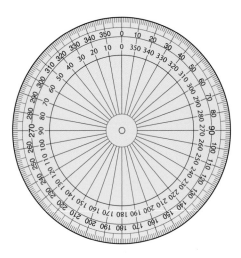

Most protractors are semicircular in shape and can be used to measure angles up to 180°.

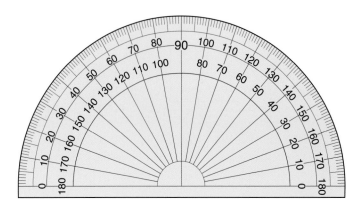

In this section you will find out how to use a semicircular protractor to measure angles, including reflex angles which are greater than 180°.

Since protractors are cheap, it is worth buying one for yourself.

EXAMPLE 2

Measure this angle.

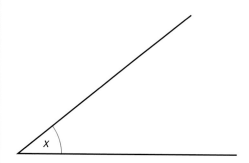

- First make an estimate.
  The angle is acute and so will be less than 90°.
  A rough estimate is about 40° since it is slightly less than half a right angle.

- Now place your protractor so that the zero line is along one of the arms of the angle.
  Make sure the centre of the protractor is at the point of the angle.

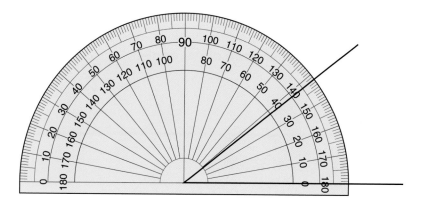

- Start at zero. Go round this scale until you reach the other arm of the angle.
  Then read the size of the angle from the scale.

Angle x = 38°

EXAM TIP

One of the most common errors in measuring angles is to use the wrong scale of the two on the protractor. Make sure you choose the one that starts at zero. A useful further check is to estimate the angle first. If the angle is obviously acute (less than $\frac{1}{4}$ turn), then the angle is less than 90°.

If the angle is obviously obtuse (greater than $\frac{1}{4}$ turn but less than $\frac{1}{2}$ turn), then the angle is between 90° and 180°. Knowing approximately what the angle is should prevent you using the wrong scale.

STAGE
2

EXAMPLE 3

Measure angle PQR.

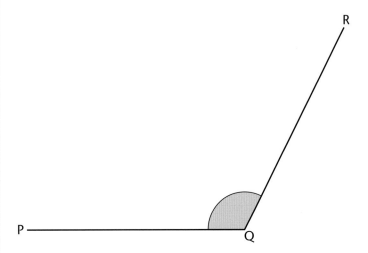

- First make an estimate.
  The angle is obtuse and so will be between 90° and 180°.
  A rough estimate is about 120°.

- Place your protractor so that the zero line is along one of the arms of the angle and the centre is at the point of the angle.

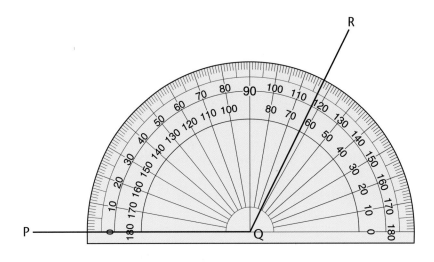

- Start at zero. Go round this scale until you reach the other arm of the angle. Then read the size of the angle from the scale.

Angle PQR = 117°

EXAMPLE 4

Measure angle A.

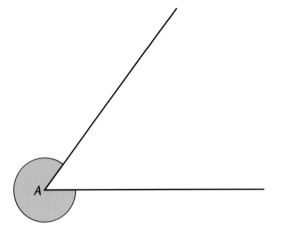

A reflex angle is between 180° and 360°.
This reflex angle is over $\frac{3}{4}$ turn so it is bigger than 270°.
A rough estimate is 300°.

You can measure an angle of this size directly using a 360° circular angle measurer. However, the scale on a semicircular protractor only goes up to 180°. You need to do a calculation as well as measure an angle.

■ Measure the acute angle first.
  The acute angle is 53°.

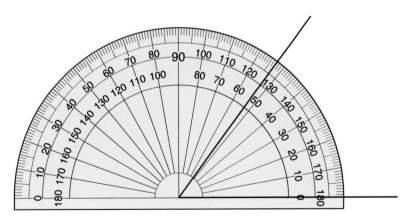

The acute angle and the reflex angle together make one full turn.
A full turn is 360°.

■ Use the fact that the two angles add up to 360° to calculate the reflex angle.
  Angle $A = 360° - 53°$
  $\qquad = 307°$

EXERCISE 3.2

1 Work out the size of each of these angles in degrees.

a) $1\frac{1}{2}$ turns   **b)** 3 turns   **c)** $\frac{1}{8}$ turn

2 Copy and complete this table by
  (i) estimating each angle.
  (ii) measuring each angle.

	Estimated angle	Measured angle
a)		
b)		
c)		
d)		
e)		
f)		
g)		
h)		
i)		
j)		

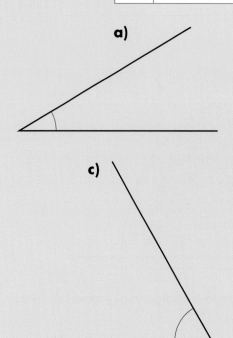

a)

c)

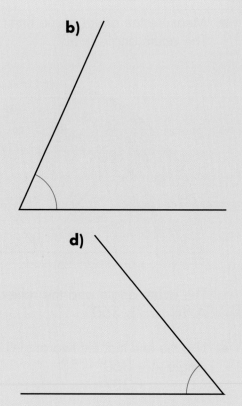

b)

d)

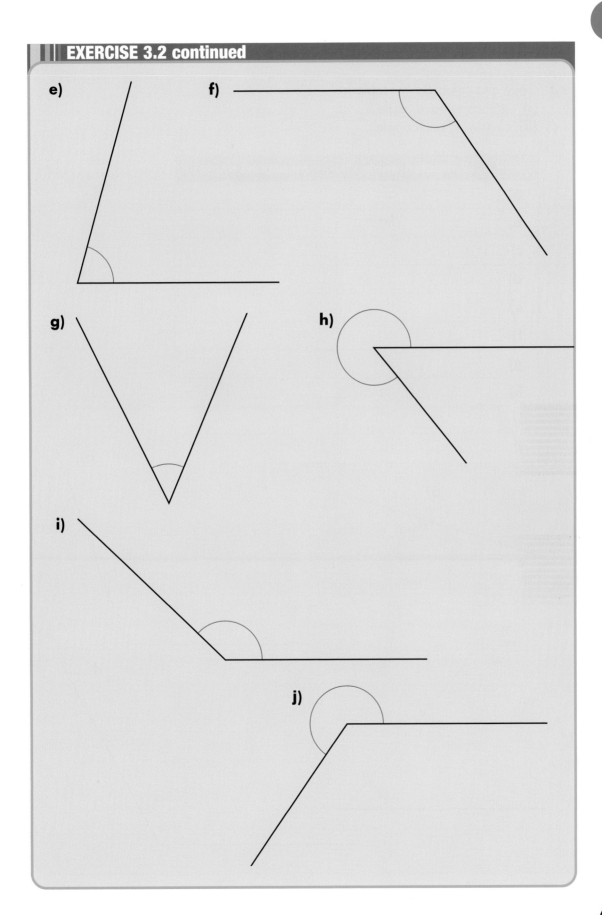

e)

f)

g)

h)

i)

j)

EXERCISE 3.2 continued

3 Copy and complete this table by

(i) estimating each angle.

(ii) measuring each angle.

	Estimated angle	Measured angle
a)		
b)		
c)		
d)		
e)		
f)		
g)		
h)		
i)		
j)		

a)

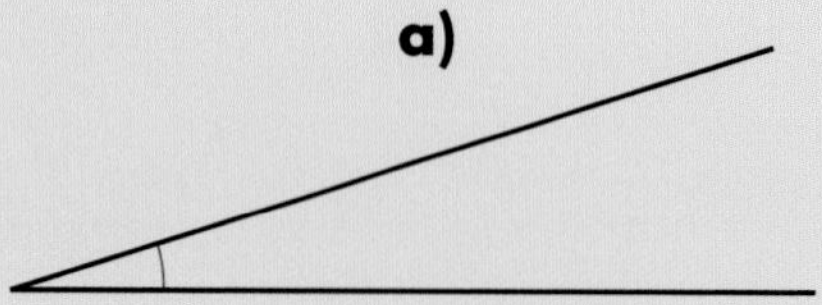

b)

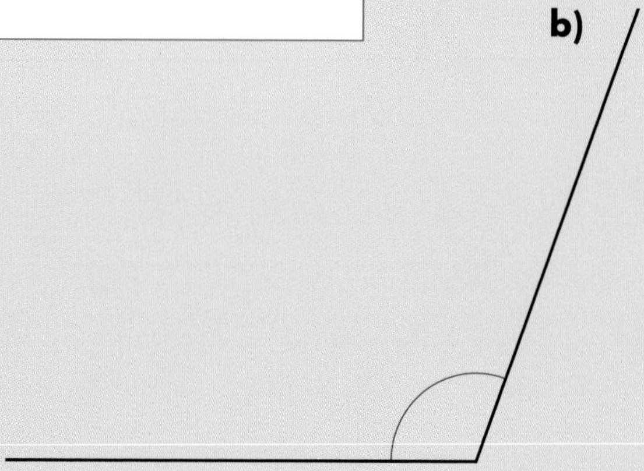

c)

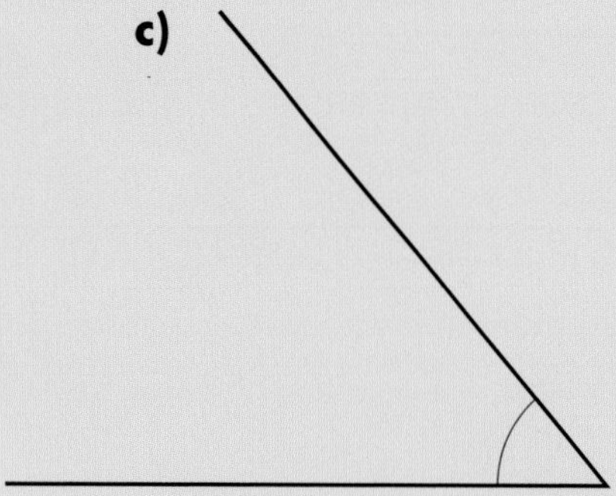

d)

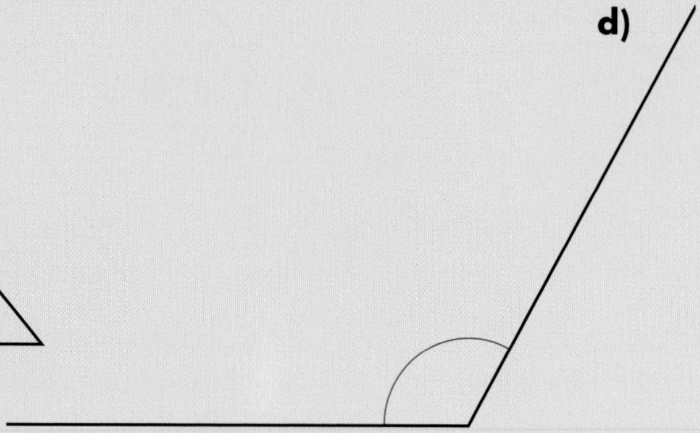

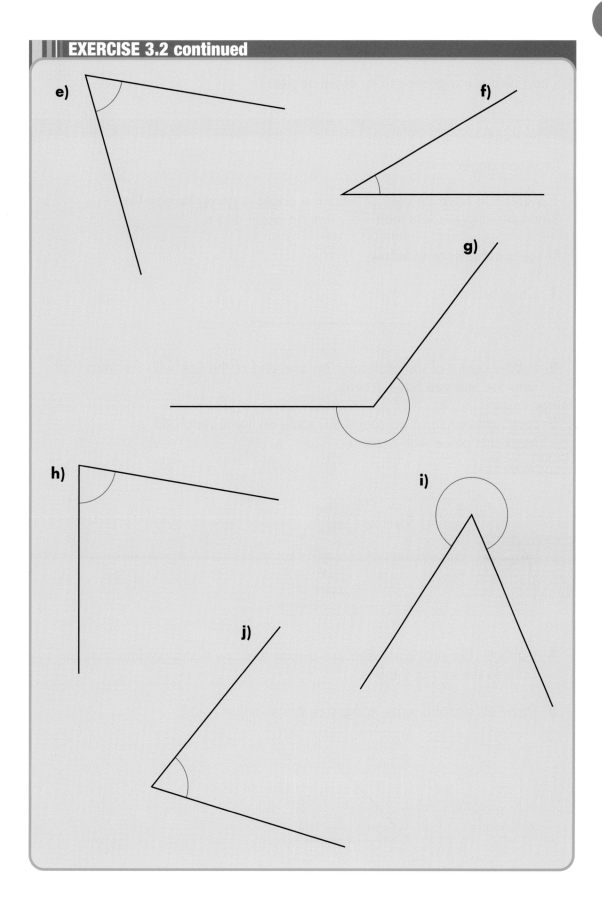

e)

f)

g)

h)

i)

j)

Drawing angles

You can also use a protractor to draw angles.

▮ EXAMPLE 5

Draw an angle of 45°.

It is useful to have an idea of what the angle is going to look like.
Since this angle is less than 90°, it is an acute angle.

These are the steps to follow.

1 Draw a line.

2 Put the centre of the protractor on one end of the line with the zero line over the line you have drawn.

3 Starting from zero, go round the scale until you reach 45°.
Mark this place with a point.

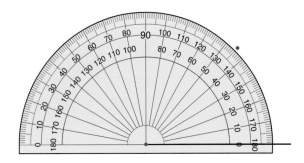

4 Remove your protractor and use a ruler to draw a straight line from the point to the end of the line.

5 Draw an arc and write in the size of the angle.

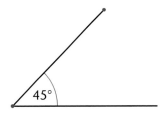

CHALLENGE 1

How would you draw an angle of 280°, a reflex angle?

Write a list of instructions like the ones in Example 5.

Hint:
You may find it useful to look back at Example 4.

EXERCISE 3.3

1 Draw accurately each of these angles.

a)
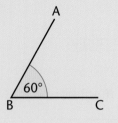

b)
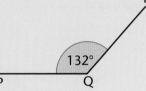

c)
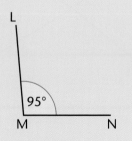

d)
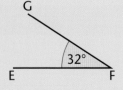

e)
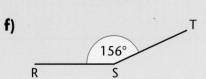

f)

2 Draw accurately each of these angles.
a) 40°   b) 90°
c) 65°   d) 27°
e) 19°   f) 38°
g) 81°   h) 73°
i) 150°   j) 116°
k) 162°   l) 98°
m) 175°   n) 144°
o) 109°   p) 127°

3 Draw accurately a reflex angle of 280°.
Follow these steps.
- First you have to do a calculation.
  360° − 280° = 80°
- Now draw this smaller angle.
- Don't forget to label the correct angle when you have finished.

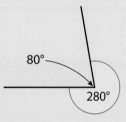

4 Use the method in question **3** to draw accurately each of these reflex angles.
a) 310°   b) 270°
c) 195°   d) 255°
e) 200°   f) 263°
g) 328°   h) 246°

C CHALLENGE 2

a) Draw a straight line.

_____

Now add another straight line.

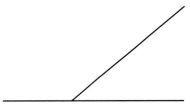

Measure the acute angle and measure the obtuse angle. Add them together.

Is the answer what you expect? Check with your neighbour.

Try again with two different angles on a straight line.

b) Draw four lines meeting at a point.

Measure the four angles and add them together.

Draw four more lines meeting at a point and measure the angles.

Do you get the same result?

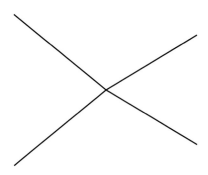

K KEY IDEAS

■ Angles are measured in degrees.

■ Acute angles are less than 90°, obtuse angles are between 90° and 180° and reflex angles are greater than 180°.

■ A full turn is 360°, a $\frac{1}{2}$ turn is 180° and a $\frac{1}{4}$ turn is 90°.

Multiplication and division

You will learn about

● Multiplying and dividing larger numbers

You should already know

● How to add and subtract numbers
● How to multiply and divide with numbers up to 10
● How to multiply and divide by 10

Multiplying larger numbers

Look at this pattern.

$1 \times 4 = 4$
$2 \times 4 = 8$
$3 \times 4 = 12$
$4 \times 4 = 16$
$5 \times 4 = 20$
$6 \times 4 = 24$
$7 \times 4 = 28$
$8 \times 4 = 32$
$9 \times 4 = 36$
$10 \times 4 = 40$

It is the 4 times table. Can you continue it?

$11 \times 4 = \square$

$12 \times 4 = \square$

The next two numbers are 44 and 48.

EXAMPLE 1

Work out 43×4.

You could continue the pattern but it would take a long time and you might make a mistake.

There are other methods you can use.

Method 1

$43 = 40 + 3$

So $43 \times 4 = 40 \times 4 + 3 \times 4$

But 40 is 10×4.

So 43×4 is the same as
$$
\begin{aligned}
10 &\times 4 \times 4 + 3 \times 4 \\
&= 10 \times 16 + 12 \\
&= 160 + 12 \\
&= 172
\end{aligned}
$$

Method 2

×	40	3
4	160	12

$$
\begin{aligned}
43 \times 4 &= 160 + 12 \\
&= 172
\end{aligned}
$$

Method 3

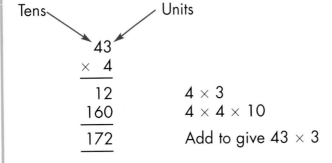

Tens   Units

$$
\begin{array}{r}
43 \\
\times\ 4 \\
\hline
12 \\
160 \\
\hline
172 \\
\hline
\end{array}
$$

4×3

$4 \times 4 \times 10$

Add to give 43×3

EXAMPLE 2

Work out 27×8.

Method 1

$$27 \times 8 = 20 \times 8 + 7 \times 8$$
$$= 10 \times 2 \times 8 + 7 \times 8$$
$$= 10 \times 16 + 56$$
$$= 160 + 56$$
$$= 216$$

Method 2

×	20	7
8	160	56

$$27 \times 8 = 160 + 56$$
$$= 216$$

Method 3

$$
\begin{array}{r}
27 \\
\times \ 8 \\
\hline
56 \\
160 \\
\hline
216 \\
\hline
\end{array}
$$

8×7

$8 \times 2 \times 10$

Add to give 27×8

EXERCISE 4.1

Work out these.

1 23×3

2 32×3

3 18×6

4 57×2

5 41×2

6 19×5

7 32×4

8 24×7

9 13×8

10 42×4

11 37×5

12 61×6

13 49×2

14 45×3

15 17×7

16 15×9

17 40×9

18 74×4

19 83 × 3	**27** 81 × 6	**35** 31 × 9
20 99 × 8	**28** 71 × 8	**36** 46 × 2
21 38 × 3	**29** 57 × 5	**37** 17 × 8
22 74 × 4	**30** 48 × 5	**38** 18 × 6
23 19 × 9	**31** 47 × 4	**39** 84 × 5
24 29 × 7	**32** 83 × 3	**40** 75 × 5
25 64 × 2	**33** 92 × 7	
26 13 × 9	**34** 91 × 9	

Dividing larger numbers

Here is part of the 8 times table.

3 × 8 = 24
4 × 8 = 32
5 × 8 = 40
6 × 8 = 48
7 × 8 = 56

The same facts could be written like this.

24 ÷ 8 = 3
32 ÷ 8 = 4
40 ÷ 8 = 5
48 ÷ 8 = 6
56 ÷ 8 = 7

STAGE
2

EXAMPLE 3

Work out $72 \div 8$.

As $9 \times 8 = 72$, $72 \div 8 = 9$.

EXAMPLE 4

Work out $72 \div 4$.

Your 4 times table probably stopped at $10 \times 4 = 40$ – unless you used the long method in Example 1!

But $10 \times 4 = 40$ gives a clue to solving the problem.

$40 \div 4 = 10$

Subtract 40 from 72, leaving 32.

$32 \div 4 = 8$

So $\quad 72 \div 4 = (40 \div 4) + (32 \div 4)$
$\qquad\qquad = 10 + 8$
$\qquad\qquad = 18$

EXAMPLE 5

Work out $72 \div 3$.

Try the same method.

$72 - 30 = 42$

42 is bigger than $3 \times 10 = 30$, so subtract another 30.

$42 - 30 = 12$

So $\quad 72 \div 3 = (30 \div 3) + (30 \div 3) + (12 \div 3)$
$\qquad\qquad = 10 + 10 + 4$
$\qquad\qquad = 24$

Multiplication and division

EXERCISE 4.2

Work out these.

1 64 ÷ 8		**11** 81 ÷ 9	
2 56 ÷ 7		**12** 99 ÷ 9	
3 64 ÷ 4		**13** 76 ÷ 4	
4 48 ÷ 3		**14** 96 ÷ 6	
5 63 ÷ 7		**15** 75 ÷ 5	
6 90 ÷ 5		**16** 84 ÷ 3	
7 39 ÷ 3		**17** 81 ÷ 3	
8 42 ÷ 2		**18** 91 ÷ 7	
9 46 ÷ 2		**19** 90 ÷ 6	
10 84 ÷ 4		**20** 56 ÷ 4	

Multiplication and division problems

A ACTIVITY 1

Work out these.

72 ÷ 3   75 ÷ 3   78 ÷ 3   81 ÷ 3

You should have the answers 24, 25, 26, 27.

What about 73 ÷ 3 and 74 ÷ 3?

73 is bigger than 72, so the answer should be bigger than 24.

But 73 is smaller than 75, so the answer should be smaller than 25.

Now try 76 ÷ 3, 77 ÷ 3, 79 ÷ 3 and 80 ÷ 3.

EXAMPLE 6

18 people are going on a journey in taxis.

Each taxi can take 4 people.

How many taxis will be needed?

To find how many taxis, work out $18 \div 4$.

$16 \div 4 = 4$ and $20 \div 4 = 5$, so the answer is between 4 and 5.

But if there are only 4 taxis, 2 people will be left behind.

So 5 taxis are needed.

EXAMPLE 7

Three friends buy a bag of sweets.

There are 47 sweets in the bag.

They share them out.

How many will each have?

To find each share, work out $47 \div 3$.

$47 = 30 + 17$
$17 = 15 + 2$

So $47 \div 3 = (30 \div 3) + (15 \div 3)$ and 2 over
$= 10 + 5$ and 2 over
$= 15$ and 2 over

So each friend has 15 sweets, and 2 sweets are left over.

The amount left over is called the **remainder**.

EXAMPLE 8

Find the remainder from each of these divisions.

a) $47 \div 4$

b) $88 \div 6$

a) $47 = 40 + 4 + 3$
$47 \div 4 = (40 \div 4) + (4 \div 4) + 3 \text{ over}$
$= 10 + 1 \text{ with remainder } 3$
$= 11 \text{ remainder } 3$

b) $88 = 60 + 24 + 4$
$88 \div 6 = (60 \div 6) + (24 \div 6) + 4 \text{ over}$
$= 10 + 4 \text{ with remainder } 4$
$= 14 \text{ remainder } 4$

If you did the calculations in Examples 6 and 7 on your calculator it would give you an answer with numbers after the decimal point.

In Example 6 your calculator would show 4·5.

In Example 7 your calculator would show 15·66… .

The numbers after the decimal point are not the same as the remainder.

To interpret the result you simply ignore any numbers after the decimal point.

As before, you need to decide whether to go to the next number.

EXERCISE 4.3

 Do not use your calculator for questions **1** to **12**.

1 Work out these.
a) 16 ÷ 3   **b)** 57 ÷ 6
c) 29 ÷ 2   **d)** 84 ÷ 9
e) 77 ÷ 4   **f)** 27 ÷ 4
g) 64 ÷ 7   **h)** 51 ÷ 2
i) 49 ÷ 6   **j)** 83 ÷ 5

2 £100 is shared between 7 people.
How many whole pounds will each get?
How much is left over?

3 A test paper has 12 pages.
How many pages are there in 9 of these papers?

4 A minibus can carry 9 people.
80 people are going on a trip.
How many minibuses will be needed?

5 I have 90 books to take upstairs.
I can only carry 8 at a time.
How many times must I go upstairs?

6 Amy plants 7 rows of bean seeds.
She plants 14 beans in each row.
How many seeds does she plant altogether?

7 Mary is packing 40 lamps in boxes.
Each box holds 6 lamps.
How many boxes can she fill?
How many lamps will be left?

8 There are 5 large mushrooms in a pack.
A greengrocer buys 19 packs.
How many mushrooms are there in total?

9 There are 88 people at a meeting.
There are 9 chairs in each row.
How many rows are needed?
How many rows will be full?

10 Five friends share a prize of £65.
How much will each get?

11 In a sale CDs cost £7 each.
Charlie buys 12 of these CDs.
How much does he spend?

12 Penny is catering for a party of people on a short holiday.
She estimates that each person will eat 9 bread rolls.
a) There are 35 people in the party.
How many rolls does she need?
b) The rolls come in packets of 6.
How many packets must she buy?

 You may use your calculator for questions **13** and **14**.

13 A printer produces 3500 copies of a book.
The books are packed in pallets of 160.
How many pallets are needed?

14 For a wedding buffet a catering company makes 6 different types of canapé.
a) They make 160 of each type.
How many canapés do they make altogether?
b) They put 18 canapés on to each plate.
How many plates do they need?

STAGE
2

189

4

C CHALLENGE 1

An earth-mover is used to carry 95 tonnes. It makes four trips.

How much does it carry each time?

C CHALLENGE 2

Look back at Activity 1.

Work out 76 ÷ 3, 77 ÷ 3, 79 ÷ 3 and 80 ÷ 3 using your calculator.

What do you notice?

Can you find the remainders?

K KEY IDEAS

■ To multiply a 2-digit number, split it into tens and units and multiply separately.

■ To divide a number larger than in the times tables, subtract 10 times the dividing number as many times as you can.

■ If there is a remainder when you divide, decide whether you need to go to the next number above.

Revision exercise A1

1 Work out these.
   a) £3 – £2·15
   b) £5 + £3·71
   c) £3·42 – £2·18
   d) 5·82 m + 6·4 m
   e) 7·14 m + 2300 m
   f) 9·4 m – 4 m 7 cm

2 **a)** Explain how you find the next number of this sequence and give the next number.

   83   78   73   68   ...

   b) Find the missing numbers in this sequence.

   ☐   15   24   ☐   ☐   51

3 Explain how you find the next calculation in this pattern and give the next calculation.

   7 × 9 + 37 = 100
   7 × 19 + 67 = 200
   7 × 29 + 97 = 300
   7 × 39 + 127 = 400

4 How many degrees are there in
   a) two complete turns?
   b) $\frac{1}{2}$ turn?

5 Measure each of these angles.

a)

b)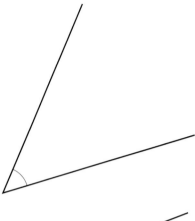

c)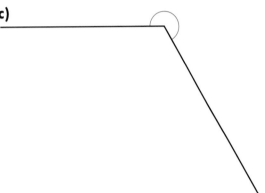

d)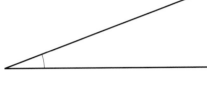

6 Draw accurately each of these angles.
  a) 50°
  b) 72°
  c) 150°
  d) 245°

7 Work out these.
  a) 73 × 4
  b) 54 × 8
  c) 16 × 9

8 Work out these.
  a) 82 ÷ 2
  b) 84 ÷ 6
  c) 77 ÷ 8

9 There are 7 classes in one year group at a school.
Each class should have 28 students.
How many students should there be in the year group?

10 How many tins, each 6 cm wide, can fit in a single line on a shelf 100 cm long?

Mass or weight?

You will learn about

● Using grams and kilograms

You should already know

● How to add and subtract decimals
● How to multiply and divide by 1000

Strictly speaking, the proper name for how heavy something is **mass**, but **weight** is the word commonly used.

Mass can be measured either in **imperial** (old) units – stones, pounds and ounces – or in **metric** units – grams and kilograms.

The imperial units for mass are rarely used now. You will sometimes see ounces used in recipes, pounds for amounts of fruit and vegetables, and stones for people's weights.

1 stone = 14 pounds and 1 pound = 16 ounces but you do not need to learn these facts.

Nearly all masses are now given in metric units.

To have a good idea of a kilogram think of a bag of sugar, which is 1 kilogram.

Small masses, like the sugar on a spoon, are weighed in grams and all larger weights are weighed in kilograms.

You do not need to write out the words in full each time.

These are the accepted short versions.

kilogram = kg   gram = g

The connection between these units is

1 kilogram = 1000 grams

(just like kilometres and metres) and you may need to change from one to the other.

EXAMPLE 1

Change each of these amounts in kilograms to grams.

a) 4 kg

b) 4·521 kg

c) 9·17 kg

As there are 1000 grams in a kilogram, to change units you just multiply by 1000.

This means that you need to move the digits three places to the left and add zeros as placeholders if necessary.

a) 4 × 1000 = 4000 g   A whole number, so add 3 zeros.

b) 4·521 × 1000 = 4521 g   The digits have moved 3 places.

c) 9·17 × 1000 = 9170 g   To move the digits 3 places you need to add a zero.

EXAMPLE 2

Change each of these amounts in grams to kilograms.

a) 7000 g

b) 4215 g

c) 82 034 g

a) 7000 ÷ 1000 = 7 kg   1 kg = 1000g, so divide by 1000, which moves the digits 3 places to the right.

b) 4215 ÷ 1000 = 4·215 kg   The same as in **a)**.

c) 82 034 ÷ 1000 = 82·034 kg

STAGE

2

5

EXAMPLE 3

Michael is baking some cakes.

One recipe needs 1·6 kg of flour.

Another needs $\frac{1}{2}$ kg of flour.

a) How much flour does he need altogether?

b) He has a new 3 kg bag of flour.
How much will be left after he has made the cakes?

a) When you have to add fractions and/or decimal parts of a kilogram, it is usually easier to change all the weights to grams.

$\frac{1}{2}$ kg = 500 g
1·6 kg = 1600 g

Total = 500 g + 1600 g
      = 2100 g
      = 2 kg 100 g or 2·1 kg

b) You can work in grams or kilograms.

In grams,

3 kg = 3000 g

Amount of flour left = 3000 g – 2100 g
                     = 900 g

In kilograms,

Amount of flour left = 3 kg – 2·1 kg
                     = 0·9 kg

STAGE
2

EXERCISE 5.1

1 What metric unit would you use to measure the mass of each of these?
  a) Yourself
  b) A bicycle
  c) A toffee
  d) An orange
  e) A £1 coin
  f) A cow
  g) An exercise book
  h) A packet of washing powder

2 Change each of these masses to grams.
  a) 9 kg
  b) 1·129 kg
  c) 3·1 kg
  d) 0·3 kg
  e) 0·012 kg
  f) 7 kg
  g) 1·13 kg
  h) 2·14 kg
  i) 0·71 kg
  j) 0·001 kg

3 Change each of these masses to kilograms.
  a) 2000 g
  b) 1400 g
  c) 3516 g
  d) 94 652 g
  e) 6600 g
  f) 8000 g
  g) 6300 g
  h) 5126 g
  i) 49 612 g
  j) 760 g

4 Write each set of masses in order of size, lightest first.
  a) 4000 g, 52 000 g, 9·4 kg, 874 g, 1·7 kg
  b) 4123 g, 2104 g, 3·4 kg, 0·174 kg, 2·79 kg

5 Some types of kitchen scales use weights which are placed on balance pans.

You have weights of these masses.
5 g, 10 g, 20 g, 20 g, 50 g, 100 g, 200 g, 200 g, 500 g, 1 kg, 2 kg

Which of these weights would you use to weigh out (i.e. balance) each of these amounts?
  a) 125 g
  b) 560 g
  c) 1·285 kg
  d) 2090 g
  e) 2·81 kg

STAGE
2

197

5

6 What is $\frac{1}{2}$ kg in grams?

7 John buys a bag of sugar weighing 1 kg, a bag of flour weighing 1·5 kg, a box of breakfast cereal weighing 450 g and two tins of soup weighing 400 g each. How much weight will he have to carry home?

8 Hannah is making a casserole.

Butter	50 g
Bacon	125 g
Onions	$\frac{1}{4}$ kg
Mushrooms	$\frac{1}{4}$ kg

What is the total weight of her ingredients?
Give your answer in kilograms.

9 Nasrin uses 750 g of sugar from a $1\frac{1}{2}$ kg bag.
How much is left?

10 A baby's weight was recorded as 4 kg 350 g.
At the next check-up, the baby's weight was recorded as 7·15 kg.
How much weight had the baby put on?

C CHALLENGE 1

If you ask an older person their weight, they will often give it in stones, or in stones and pounds.

■ This man weighs 12 stone and 4 pounds.

Find out how much that is in kilograms.

■ How much do you weigh in stones?

K | **KEY IDEAS**

- Everyday units of mass are kilograms (kg) and grams (g).

- There are 1000 grams in a kilogram.

6 Probability

You will learn about

- Using a probability scale
- Probabilities being numbers between 0 and 1

You should already know

- Words used to decribe the probability of an event
- How to read scales

Probability scales

In Stage 1 you learnt that you can use words such as 'certain' and 'evens' to describe the probability of an event happening.

You can place these words on a probability scale, like this. At one end of the scale you place 'impossible' and at the other end you place 'certain'.

| impossible | very unlikely | unlikely | evens | likely | very likely | certain |

EXAMPLE 1

Here is a probability scale.

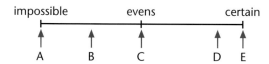

Match each of these events with a letter on the probability scale.

a) If you toss a fair coin it will come down tails.

b) If you roll an ordinary fair dice it will give a 7.

c) Christmas Day will fall on 25 December this year.

d) If you take a card from a pack of ordinary playing cards it will be a heart.

e) In England it will rain sometime during the next 14 days.

a) The chance that a fair coin will come down tails is evens, so this event matches with C.

b) You cannot get a 7 on a normal dice, so this event is impossible and matches with A.

c) It is certain that Christmas Day will be on 25 December this year, so this event matches with E.

d) There are quite a lot of hearts in a pack of cards but fewer than half the cards are hearts, so this event is between evens and impossible and matches with B.

e) It is very likely that it will rain in England sometime during the next 14 days, so this event is close to certain and matches with D.

EXERCISE 6.1

For each question make a copy of this probability scale.

impossible   evens   certain

1 Place and label arrows on your scale to show the probability of each of these events.
  a) Getting an odd number when you roll an ordinary fair dice.
  b) It will get dark tonight.
  c) You will swim the Channel.

2 Place and label arrows on your scale to show the probability of each of these events.
  a) You will win the National Lottery.
  b) Getting a liquorice sweet in when you take a sweet from a bag of Liquorice Allsorts.
  c) Getting a number greater than 4 when you roll an ordinary six-sided dice.

3 Place and label arrows on your scale to show the probability of each of these events when you pick a letter from the word CHOCOLATES.
  a) Getting a vowel.
  b) Getting a letter before M in the alphabet.
  c) Getting a letter M.

4 Place and label arrows on your scale to show the probability of each of these events.
  a) It will rain every day in April.
  b) It will snow in the Sahara Desert next August.
  c) The River Thames will freeze next week.

5 Place and label arrows on your scale to show the probability of each of these events.
  a) There will be a general election within the next five years.
  b) It will snow in August.
  c) Getting a 2 when you roll an ordinary six-sided dice.

6 You take one playing card without looking from these five cards.

Draw arrows on your probability scale to show the probability of
  a) taking a red card.
  b) taking a black card.

7 You have a bag containing five red and five blue balls. Draw arrows on your probability scale to show the probability of
  a) picking a red ball without looking.
  b) picking a yellow ball without looking.

8 Place and label arrows on your scale to show the probability of each of these events when you take one card from an ordinary pack of playing cards.
  a) Getting a king, a queen or a jack.
  b) Getting a card with a number less than 10.
  c) Getting a blank card.

Using numbers for probabilities

Probability uses numbers to show how likely an event is.

The probability of any event happening must lie between 0 and 1.

- 0 is the probability of an event which cannot happen.

- 1 is the probability of an event which is certain to happen.

- 0.5 or $\frac{1}{2}$ is the probability of an event that has an evens chance of happening.

You can add these numbers to a probability scale, like this.

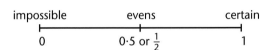

Often the scale is shown with just the numbers, like this.

```
0              0·5 or ½              1
├──────────────┼──────────────┤
```

EXAMPLE 2

Place and label arrows on this probability scale to show the probability of each of these events.

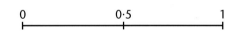

a) The Atlantic Ocean will freeze tomorrow.
b) The sun will shine on 1 June.
c) You will get a number less than 3 when you roll an ordinary six-sided dice.

a) It is impossible that the Atlantic Ocean will freeze tomorrow so you place an arrow at zero.

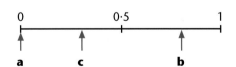

b) It is likely that the sun will shine on 1 June, but not certain, so you place an arrow between evens, which is a probability of 0·5, and certain, which is a probability of 1.

c) A number less than 3 is 1 or 2. This is unlikely but not very unlikely and not impossible so you place an arrow between impossible and evens, closer to evens.

STAGE
2

EXERCISE 6.2

For each question make a copy of this probability scale.

```
0              0·5              1
├───────────────┼───────────────┤
```

Draw arrows on your scale to show the probability of each of the events.

1 Show the probability of each of these events.
   a) Choosing a chocolate eclair from a bag of mint humbugs.
   b) Without looking, taking out a red counter from a bag of ten red counters and two blue counters.
   c) You will have a drink tomorrow.

2 Show the probability of each of these events.
   a) Getting an even number when you roll an ordinary six-sided dice.
   b) Getting a spade or a club card when you take one card from a pack of ordinary playing cards.
   c) Getting a head or a tail when you spin an ordinary coin.

3 Show the probability of each of these events.
   a) There will be a gale tomorrow.
   b) Getting a number greater than 10 when you roll an ordinary six-sided dice.
   c) Getting a heart card when you take one card from a pack of ordinary playing cards.

4 Show the probability of each of these events.
   a) Getting a number greater than 2 when you roll an ordinary six-sided dice.
   b) A new car will break down during the first year.
   c) You will experience an earthquake tomorrow.

5 Show the probability of each of these events.
   a) A coin landing and remaining on its edge when you toss it.
   b) You will fly to the moon before you are 20.
   c) Without looking, taking an ace from a pack of ordinary playing cards.

C   CHALLENGE 1

You have an ordinary six-sided dice.

What is the probability that you will get a 6 when you roll it?

Try this experiment.

- Roll the dice 100 times (remember to count).

- Record the number of 6s.

Can you answer the question now?

K   KEY IDEAS

- The probability of an event happening can be shown on a probability scale.

- Probability is a number and shows how likely an event is.

- The probability of any event happening lies between 0 and 1.
  - 0 is the probability of an event which cannot happen.
  - 1 is the probability of an event which is certain to happen.
  - 0·5 or $\frac{1}{2}$ is the probability of an event that has an evens chance of happening.

Probability

STAGE
2

205

Percentages, fractions and decimals

You will learn about

- Percentages
- Percentage scales and pie charts
- Simple fractions and decimals

You should already know

- How to find $\frac{1}{4}$, $\frac{1}{2}$ and $\frac{3}{4}$ of an amount
- How to multiply and divide using a calculator
- How to read a scale

Percentages

'Per cent' means 'out of 100'. The % symbol is used as a short way of writing 'per cent'.

Look at this statement.

> 50% of cat owners say their cats prefer Frolic cat food.

50% means 50 out of 100.

So the statment is saying that 50 out of every 100 cat owners say their cats prefer Frolic.

You could show this in a 10 × 10 square like this.

You can see that 50% is the same as $\frac{1}{2}$.

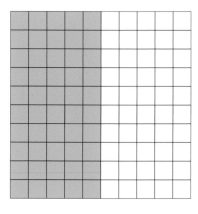

Here is another statement using a percentage.

25% of dog owners feed their dogs with dry food.

25% means 25 out of 100.

Again, you can show this on a 10 × 10 square.

You can see that 25% is the same as $\frac{1}{4}$.

You can also see that 75% is the same as $\frac{3}{4}$.

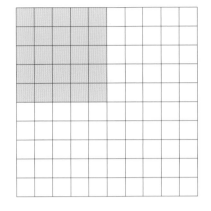

In Stage 1 you learned how to find $\frac{1}{2}$, $\frac{1}{4}$ and $\frac{3}{4}$ of an amount.

Since 50% is the same as $\frac{1}{2}$, you can find 50% of an amount by finding $\frac{1}{2}$ of that amount.

Similarly, you can find 25% of an amount by finding $\frac{1}{4}$ of the amount.

You can find $\frac{3}{4}$ of an amount by finding 50% of it and 25% of it and adding the answers together.

EXAMPLE 1

In the survey on page 206, 50% of cat owners said their cats prefer Frolic cat food.

There were 300 cat owners in the survey.

How many cat owners said their cats prefer Frolic cat food?

50% is the same as $\frac{1}{2}$.

To find $\frac{1}{2}$ of an amount you divide by 2.

300 ÷ 2 = 150

7

|| EXAMPLE 2

In the survey on page 207, 25% of dog owners said they feed their dogs dry food.

There were 200 dog owners in the survey.

How many dog owners said they feed their dogs dry food?

25% is the same as $\frac{1}{4}$.

To find $\frac{1}{4}$ of an amount you divide by 4.

200 ÷ 4 = 50

|| EXERCISE 7.1

1 Work out 25% of each of these amounts.
    a) £12   **b)** £20   **c)** £4   **d)** £8·40

2 Find 50% of each of these amounts.
    a) £10   **b)** 120 cm   **c)** 20 kg   **d)** £1500

3 Find 75% of each of these amounts.
    a) 48 kg   **b)** £16   **c)** £60   **d)** 24 m

4 Eddie used to get £4·50 each week for cleaning cars.
    He was given a 50% pay rise.
    a) How much extra did he get?
    b) What was his new wage?

5 'Books R Us' sell books by post.
    You choose one or more books each month, but pay 25% of the cost for post and packing.
    a) Jodi buys some books that cost £20.
        How much will the post and packing be?
    b) How much will it cost her altogether?
    c) Rachel buys three books which cost £8 each.
        How much will she have to pay altogether?

STAGE
2

EXERCISE 7.1 continued

6 In a clearance sale, Stars Electrical Discount Stores offer these items with a 25% reduction.
Copy and complete the table.

Item	Normal price	Reduction	Sale price
Washing machine	£400		
TV	£150		
Fridge-freezer	£240		
Microwave	£100		

7 A test was marked out of 60.
John gained 50% and Susie gained 25%.
How many marks did each get?

8 A packet of rice states that it contains 25% extra, and gives the usual mass as 300 g.
How much extra rice is in the packet?

9 The height of a retaining wall is 50% of its length.
What is the height of the wall if the length is 5 m?

Percentage scales

A

ACTIVITY 1

Look at this scale. The left-hand end is 0% and the right-hand end is 100%.

0%   100%

Make a copy of the scale and mark in the values on all the lines in between.

EXAMPLE 3

This scale shows what percentage of Northern Branch line trains were on time last year.

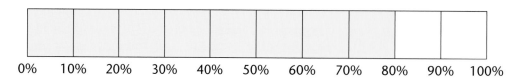

0%　10%　20%　30%　40%　50%　60%　70%　80%　90%　100%

a) What percentage were on time?

b) What percentage were not on time?

a) Each division is 10% and eight divisions are shaded, so 80% of trains were on time.

b) Two divisions are unshaded, so 20% were not on time.

> **EXAM TIP**
> The scale does not show how many trains there were, only what percentage.

Percentage pie charts

Instead of using a straight line, we can use a circle. This 'pie' is divided into ten slices, called **sectors**. Each sector is 10% of the whole pie.

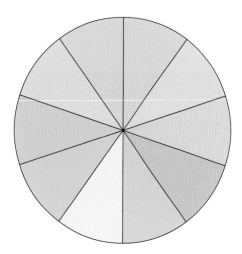

STAGE

2

Pie charts show how quantities are shared out. This one could show how a lottery prize is shared between ten holders of a winning ticket.

Like other percentage scales, pie charts do not show what the total is.

Sometimes pie charts are marked to show percentages like this.

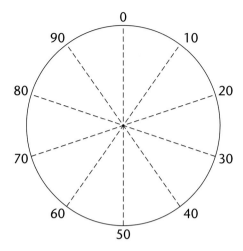

About 70% of the Earth's surface is water and 30% is land.

This can be shown in a pie chart.

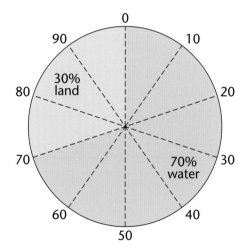

You can see that the blue sector for water is from 0% to 70%.

Counting the other way from 0% for the green sector (land), there are three 10% sectors.

EXERCISE 7.2

1 This scale shows what percentage of Camtrak trains were on time last year.

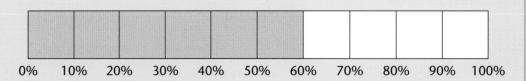

a) What percentage were on time?
b) What percentage were not on time?
c) Look back at the scale in Example 3.
Which trains were more often on time, Camtrak or Northern Branch line?
What was the difference?

2 This pie chart shows how Kevin
spends a typical day.
In parts **a)** and **b)** choose the
answer you think is correct.
a) The percentage of time spent
watching TV is
(i) more than 25%.
(ii) 25%.
(iii) less than 25%.
b) The percentage of time spent
sleeping is
(i) more than 50%.
(ii) less than 50%.
(iii) 50%.
c) Find the actual percentages.

3 This percentage scale shows how Perry spends a typical day.
Find what percentage of time Perry spends on each activity.

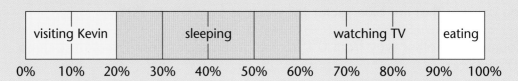

4 Joanne asked people which TV channel they watched between 10 p.m. and midnight one Friday night.
She showed her results in a pie chart.

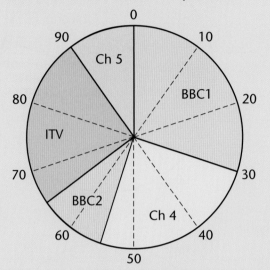

EXAM TIP
Always work from a 10% line if you can.

Find the percentage of people who watched each channel.

5 These percentage scales show how two people spent their money.

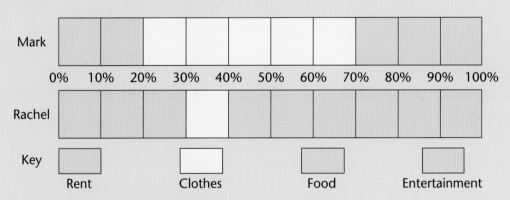

a) What percentage of her money did Rachel spend on clothes?
b) Who spent the larger percentage on entertainment?
c) What did Mark spend half his money on?

STAGE
2

6 A school collected data on where its students went when they left Year 11. The pie chart shows the results.

What percentage of students followed each course?

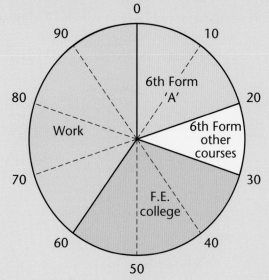

7 This pie chart shows the reading habits of a group of people.

Estimate what percentage of people read each newspaper.

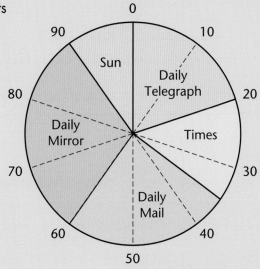

8 A class of students were asked what sort of pet they had. The pie chart shows the results of the survey.

Estimate what percentage of students had each kind of pet.

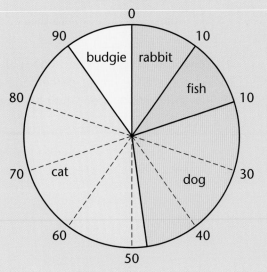

CHALLENGE 1

Look back at question **5**.

Can you say who spent more on clothes?

Can you say who earns more money?

Explain your answers.

CHALLENGE 2

The Earth has four great oceans – the Pacific, Atlantic, Indian and Arctic Oceans.

Approximately, they represent these percentages of the total surface area of the oceans.

Ocean	Percentage
Pacific	50
Atlantic	25
Indian	20
Arctic	5

Draw a pie chart to show this information.

STAGE

2

Fractions and decimals

In Stage 1 you learnt that you can use fractions to say how many parts you have of a whole.

For example, this shape is divided into five equal parts and two of them are blue.

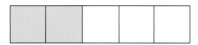

You can say that $\frac{2}{5}$ is blue.

A ACTIVITY 2

You have already found the connection between $\frac{1}{4}$, $\frac{1}{2}$, $\frac{3}{4}$ and 25%, 50%, 75%.

Taking half of a quantity means dividing by 2.

Use your calculator to check that $1 \div 2 = 0.5$.

Now do the same for $\frac{1}{4}$ and $\frac{3}{4}$.

Other fractions can be turned into decimals in the same way.

Work out the decimal for each of these fractions.

$\frac{1}{5} =$ $\frac{2}{5} =$ $\frac{3}{5} =$ $\frac{4}{5} =$

$\frac{1}{10} =$ $\frac{2}{10} =$ $\frac{3}{10} =$ $\frac{4}{10} =$ $\frac{5}{10} =$ $\frac{6}{10} =$ $\frac{7}{10} =$ $\frac{8}{10} =$ $\frac{9}{10}$

EXAM TIP
You will need to remember the decimal equivalents of these fractions.

C CHALLENGE 3

Now try these fractions.

$\frac{1}{3} =$ $\frac{2}{3} =$ $\frac{1}{9} =$ $\frac{8}{9} =$ $\frac{2}{7} =$

What happens when you turn them into decimals?

||| EXERCISE 7.3

1 What fraction of each of these shapes is blue?

a)

b)

c)

d)

e)

f)

g)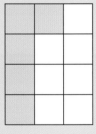

h)

2 What fraction of each shape in question **1** is not blue?

3 What fraction of each of these shapes is blue?

a) 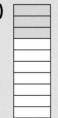   b)   c)   d)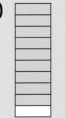

4 Turn each of the fractions in question **3** into a decimal.

Percentages, fractions and decimals

5 On squared paper, make four copies of this shape.

Shade some of the squares to represent each of these fractions.

a) $\frac{7}{10}$ **b)** $\frac{1}{2}$ **c)** $\frac{2}{5}$ **d)** $\frac{4}{5}$

6 Turn each of the fractions in question **5** into a decimal.

7 Draw a suitable shape and shade it to represent each of these fractions.

a) $\frac{3}{4}$ **b)** $\frac{1}{10}$ **c)** $\frac{3}{5}$ **d)** $\frac{4}{10}$

8 Turn each of the fractions in question **7** into a decimal.

9 Copy and complete this table.

Fraction	Decimal	Percentage
		25%
	0·5	
$\frac{3}{4}$		

K **KEY IDEAS**

■ To work out simple percentages, use the fraction which is the same, e.g. for 25% use $\frac{1}{4}$.

■ Percentages on a percentage scale or pie chart can be read from 10% divisions.

■ You need to know these fraction to decimal equivalents.

$\frac{1}{2} = 0.5$ $\frac{1}{4} = 0.25$ $\frac{3}{4} = 0.75$

$\frac{1}{5} = 0.2$ $\frac{2}{5} = 0.4$ $\frac{3}{5} = 0.6$ $\frac{4}{5} = 0.8$

$\frac{1}{10} = 0.1$ $\frac{2}{10} = 0.2$ $\frac{3}{10} = 0.3$ $\frac{4}{10} = 0.4$ $\frac{5}{10} = 0.5$

$\frac{6}{10} = 0.6$ $\frac{7}{10} = 0.7$ $\frac{8}{10} = 0.8$ $\frac{9}{10} = 0.9$

STAGE
2

Negative numbers

You should already know

- How to add and subtract whole numbers
- How to read whole numbers off a scale

Some numbers are less than zero. These are called **negative numbers**. They are written as ordinary numbers with a minus sign in front.

The minus sign tells you that the number is below zero and the number tells you how far it is below zero.

Negative numbers are used in many situations.

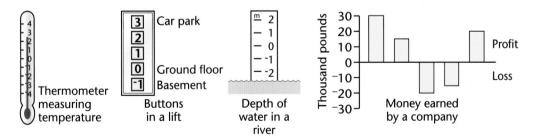

In this chapter you will be concentrating on the use of negative numbers on the temperature scale.

This is a temperature scale from ⁻10° to 10°.

Notice that ⁻10° is lower than ⁻1°.

⁻10° is 10° below zero but ⁻1° is only 1° below zero.

Remember that zero is neither positive nor negative.

The temperature scale may help you with the work in this chapter. Copy it into your book. Make it longer if you wish.

A ACTIVITY 1

This table gives the average temperatures in January and in July for ten European cities.

City	January (°C)	July (°C)
Amsterdam	2	17
Athens	9	27
Berlin	⁻1	18
Bucharest	⁻3	22
Budapest	⁻1	21
Dublin	5	15
Moscow	⁻9	18
Paris	3	19
Stockholm	⁻3	18
Vienna	⁻1	19

a) Put the January temperatures on a number line.

b) Write the January temperatures in order, lowest first.

c) Work out the difference between the highest and the lowest temperatures for each of the cities.

d) Which city has
(i) the greatest difference between the highest and the lowest temperatures?
(ii) the smallest difference?

STAGE
2

EXAMPLE 1

Margaret measured the daytime and night-time temperatures in her garden for two days.

Here are her results.

Day	Monday daytime	Monday night-time	Tuesday daytime	Tuesday night-time
Temperature (°C)	7	⁻2	3	⁻5

a) How much did the temperature change between each reading?

b) The daytime temperature on Wednesday was 4°C warmer than the Tuesday night-time temperature.
What was the Wednesday daytime temperature?

Use the temperature scale to help you.

a) The Monday daytime temperature is 7°C.
The Monday night-time temperature is ⁻2°C.
To get from 7°C to ⁻2°C you go down 9°C.

The Tuesday daytime temperature is 3°C.
To get from ⁻2°C to 3°C you go up 5°C.

The Tuesday night-time temperature is ⁻5°C.
To get from 3°C to ⁻5°C you go down 8°C.

b) Find ⁻5°C on the scale and move 4°C up.
You get to ⁻1°C.

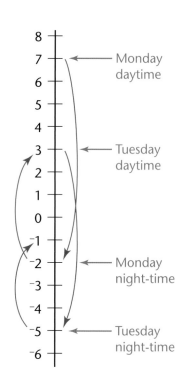

EXERCISE 8.1

1 Copy and complete these sentences.
Use the words 'warmer' or 'colder' to fill the gaps.
a) 7°C is .... than 2°C.
b) ⁻10°C is .... than 5°C.
c) ⁻3°C is .... than 0°C.
d) ⁻7°C is .... than ⁻12°C.
e) 10°C is .... than 0°C.
f) ⁻2°C is .... than ⁻5°C.
g) 4°C is .... than ⁻1°C.
h) ⁻2°C is .... than 2°C.

2 The table gives the average temperatures in January and in July for five cities.

City	January (°C)	July (°C)
Casablanca	12	22
Jeddah	23	33
New York	0	25
Montreal	⁻10	21
Beijing	⁻5	31

a) Put the January temperatures on a number line.
b) Write the January temperatures in order, lowest first.
c) Work out the difference between the highest and the lowest temperatures for each of the cities.
d) Which city has the greatest difference between the highest and the lowest temperatures?

3 Write each set of temperatures in order, lowest first.
a) ⁻2°C   7°C   0°C   ⁻5°C   3°C
b) ⁻2°C   5°C   1°C   2°C   ⁻1°C
c) 7°C   ⁻7°C   4°C   ⁻9°C   ⁻3°C
d) 9°C   4°C   ⁻2°C   7°C   ⁻8°C   ⁻1°C
e) ⁻4°C   5°C   ⁻2°C   3°C   ⁻7°C
f) ⁻8°C   2°C   8°C   0°C   ⁻9°C
g) ⁻4°C   ⁻1°C   6°C   8°C   5°C   ⁻3°C
h) ⁻10°C   8°C   ⁻5°C   ⁻9°C   2°C   ⁻1°C

STAGE

2

4 Copy the table and fill in the gaps.

	Start temperature (°C)	Move (°C)	Finish temperature (°C)
a)	2	Up 7	
b)	0	Down 6	
c)	‾6	Up 2	
d)	4		‾7
e)	‾6		‾10
f)	3		‾3
g)		Down 4	‾7
h)		Up 3	‾5
i)		Down 6	0
j)		Up 7	4

5 Copy the table and fill in the gaps.

	Start temperature (°C)	Move (°C)	Finish temperature (°C)
a)	4	Up 3	
b)	‾2	Down 4	
c)	10	Down 14	
d)	‾10		‾2
e)	10		‾9
f)	‾4		2
g)		Up 7	10
h)		Down 6	‾9
i)		Up 2	‾8
j)	‾5	Down 3	

6 Write down the next three temperatures in each of these sequences.

a) 8°   5°   2°   ...   **b)** 2°   1°   0°   ...
c) ⁻10°   ⁻8°   ⁻6°   ...   **d)** 10°   5°   0°   ...
e) 10°   6°   2°   ...   **f)** 4°   2°   0°   ...
g) ⁻9°   ⁻6°   ⁻3°   ...   **h)** ⁻40°   ⁻30°   ⁻20°   ...

7 What is the difference in temperature between each of these?

a) 17°C and ⁻1°C   **b)** ⁻8°C and 12°C
c) ⁻19°C and ⁻5°C   **d)** 30°C and ⁻18°C
e) 13°C and ⁻5°C   **f)** ⁻10°C and 15°C
g) ⁻20°C and ⁻2°C   **h)** 25°C and ⁻25°C

C CHALLENGE 1

This table gives the heights above sea level, in metres, of seven places.

Place	Height (m)
Mount Everest	8 863
Bottom of Lake Baikal	⁻1 484
Bottom of Dead Sea	⁻792
Ben Nevis	1 344
Mariana Trench	⁻11 022
Mont Blanc	4 807
World's deepest cave	⁻1 602

a) Put these heights in order, lowest first.

b) What is the difference in height between the highest and lowest places?

STAGE
2

8

K KEY IDEAS

- Negative numbers are less than zero. The minus sign tells you that the number is below zero and the digits tell you how far below zero.

- Positive and negative numbers can be represented on a number line. A number lower on the number line than another is a smaller number.

Revision exercise B1

1 Change each of these masses to the units indicated.
   a) 25·23 kg to grams
   b) 14 080 g to kilograms

2 Find the total mass of 1·43 kg, 500 g, 955 g and 23·54 kg.

3 Write these masses in order, smallest first.
3 kg, 300 g, 1·4 kg, 2 g, 875 g, 2500 g

4 Copy this scale.

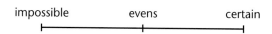

impossible   evens   certain

Place and label arrows on your scale to show the probability of each of these events.
   a) Valentine's Day will be on 14 February.
   b) The next vehicle to pass your school will be a car.
   c) It will rain during the next seven days.

5 Copy this scale.

0          $\frac{1}{2}$ or 0·5   1

Place and label arrows on your scale to show the probability of each of these events.
   a) The next person to enter a restaurant is female.
   b) Getting a 1 when you throw an ordinary dice.
   c) Getting a number larger than 6 when you throw an ordinary dice.

6 Find $\frac{1}{4}$ of 72.

7 Find these.
   a) 50% of £54
   b) 75% of 36 g

8 In a sale everything is reduced by 50%. How much will a suit originally priced at £110 sell for in the sale?

9 A courier delivers a package worth £200.
He charges a 25% handling fee. How much is the handling fee?

10 The pie chart shows the colours of cars in a car park.

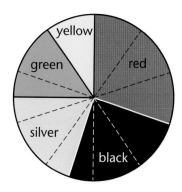

Find what percentage of the cars were each colour.

11 Turn each of these fractions into a decimal.
   a) $\frac{1}{4}$ **b)** $\frac{1}{5}$
   c) $\frac{1}{10}$ **d)** $\frac{3}{4}$
   e) $\frac{4}{5}$

12 Write these temperatures in order, lowest first.
8°C, ⁻3°C, 0°C, 5°C, ⁻1°C, ⁻6°C

13 Copy and complete these sentences. Use the words 'higher' or 'lower' to fill the gaps.

a) 8°C is a .... temperature than ⁻10°C.

b) ⁻2°C is a .... temperature than 0°C.

c) ⁻5°C is a .... temperature than ⁻9°C.

14 During one winter day, the minimum temperature was ⁻4°C and the maximum temperature was 7°C. What was the difference between these temperatures?

Estimating angles and lengths

9

You should already know

- How to identify types of angle

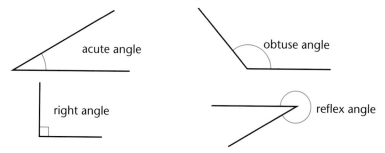

acute angle

obtuse angle

right angle

reflex angle

- How to use a protractor or angle measurer to measure angles
- How to read the scales on a ruler

9

To estimate the size of an angle or length, compare it with one you know.

EXAMPLE 1

The angle on the left measures 20°.

Estimate the size of the one on the right.

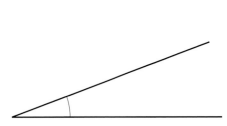 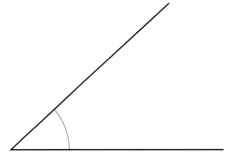

The angle on the right is about twice as big as the one on the left.

It is about 2 × 20° = 40°.

EXAMPLE 2

Estimate the size of this angle.

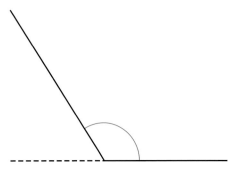

The angle is about two thirds round to a straight line, which is 180°.

It is about 120°.

EXERCISE 9.1

1 This angle measures 30°.

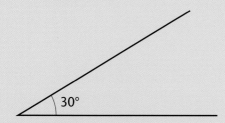

Estimate the size of each of these angles.
Write down your estimates and then check them using a protractor or angle measurer.

a)

b)

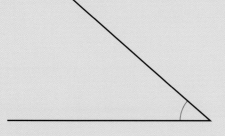

c)

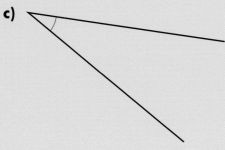

STAGE
2

2 This angle measures 60°.

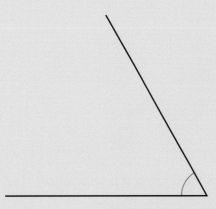

Estimate the size of each of these angles.
Write down your estimates and then check using a protractor or
angle measurer.

a)

b)

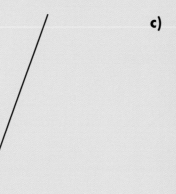

c)

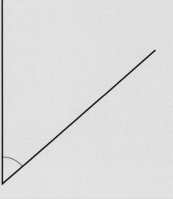

3 These angles measure 45° and 135° respectively.

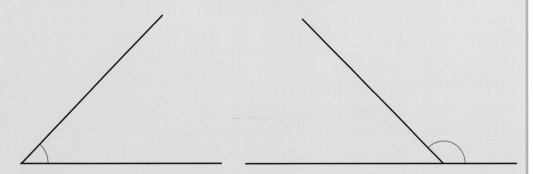

Estimate the size of each of these angles.
Write down your estimates and then check using a protractor or
angle measurer.

a)

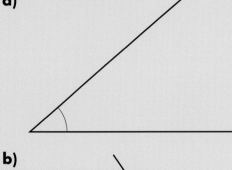

b)

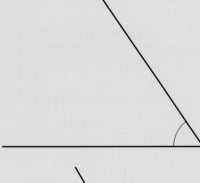

d)

c)

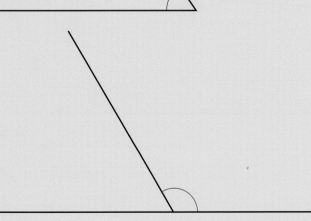

4 This angle measures 20°.

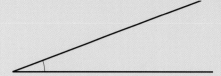

Estimate the size of each of these angles.
Write down your estimates and then check using a protractor or angle measurer.

a)

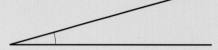

b)

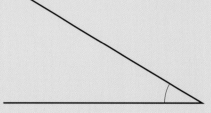

5 This line is 10 cm long.

10 cm

Estimate the length of each of these lines and then check by measuring.

a) ────────────────

b) ──────────────────────────

c) ──────────────────

6 This line is 15 cm long.

15 cm

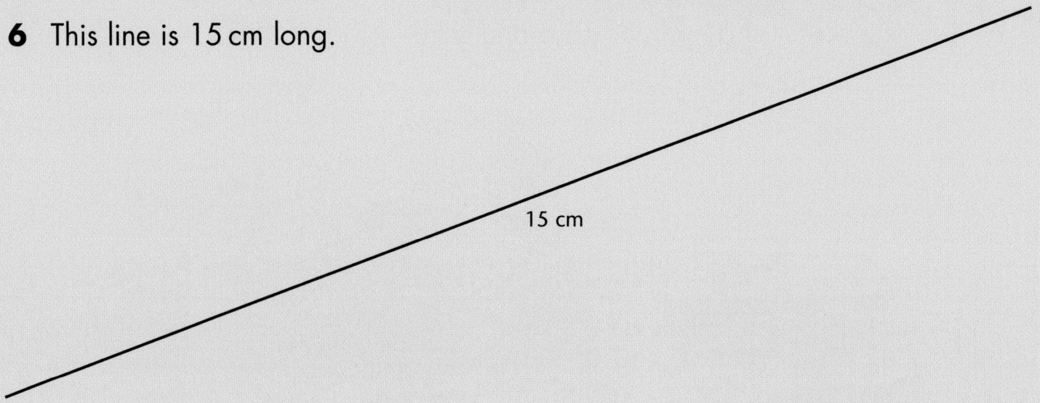

Estimate the length of each side of this rectangle.

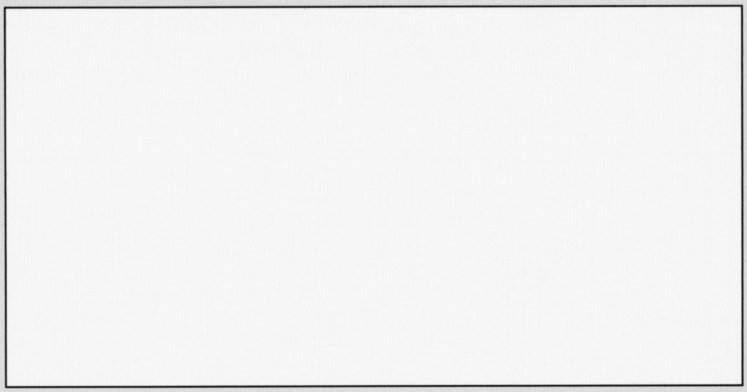

7 Estimate the length of several objects and then check by measuring.
Show your results in a table. Three suggestions are given for you.

Object	Estimated length	Measured length
Book		
Width of desk		
Pen or pencil		

STAGE

2

8 Here is a sketch of a man, a house and a tree.

The man is about 2 m tall. Estimate the height of the house and the tree.

9 This is a sketch map of a railway line.

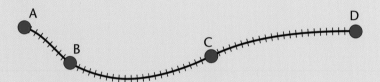

It is 20 km from A to B. Estimate the distance from B to C and from A to D.

10 The person in this picture is about 2 m tall.
Estimate the height of the tower.

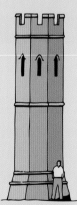

11 The man in this picture is approximately 2 m tall.
Estimate the height of the Christmas tree.

12 The car in this picture is approximately 1·5 m high.
Estimate the height of the building.

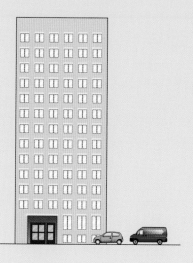

9

C CHALLENGE 1

How can you estimate longer distances such as the length of a field or the height of a tall chimney?

K KEY IDEAS

■ To estimate the size of an angle, first compare it with a right angle and decide whether it is an acute, obtuse, reflex or right angle.
Then compare it with the given angle.

■ To estimate an unknown length, compare it with a known length.

STAGE
2

Solids 10

3-D shapes

Two 3-D shapes that you may not have met before are the **prism** and the **tetrahedron**.

Here are two examples of prisms.

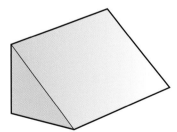

Triangular prism

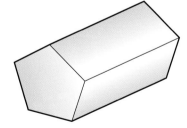

Pentagonal prism

Look at the shape of the faces.

The shape on each end of the triangular prism is a triangle.
All the other faces are rectangles.

If you cut through the triangular prism at right angles to one of the rectangular faces you will see its **cross-section**. It is the same shape as the triangle at either end of the prism.

The cross-section of a prism is the same all the way through.

The cross-section of a prism can be any shape.

The cross-section of the second prism on page 239 is a five-sided shape (pentagon).

A **tetrahedron** is a special pyramid. Its base is a triangle.

You can see that all its four faces are triangles.

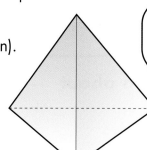

EXERCISE 10.1

1 Name each of these shapes.

a)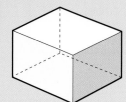

b)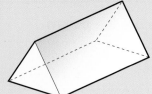

c)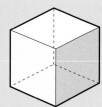

d)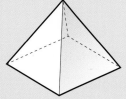

3 Name each of these shapes.

a)

b)

c)

d)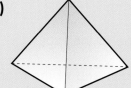

2 How many faces and how many vertices do each of the shapes in question **1** have?

4 Which of the shapes in questions **1** and **3** are also prisms?

Making 3-D shapes

A flat shape which can be folded to make a 3-D shape is called its **net**.

When actually making 3-D shapes, you need to add flaps to the edges so that they can be glued together. These flaps are called **tabs**.

ACTIVITY 1

Here is a net.

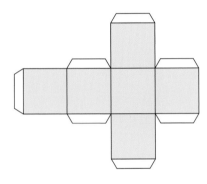

Cut out a copy of the shape with the sides of each of the squares 3 cm long.

Fold along the lines to make a 3-D shape.

What shape have you made?

> **EXAM TIP**
> Give yourself practice in selecting and describing shapes to help you remember the words needed.

Here are some tips for making 3-D shapes.

■ Put tabs on every other edge.

■ Use card rather than paper, if possible, to construct a shape that will last!

■ When using card, score the edges before folding.

■ Use glue that is suitable for the material you are using. If possible, use quick-drying glue.

■ Use the resulting shapes if you can – when suitably decorated, they can make good gift or storage boxes, for example.

Look at each of these arrangements of squares.

Which do you think will fold to make a cube?

Now cut out and fold a copy of each of the shapes to check your answers.

a)

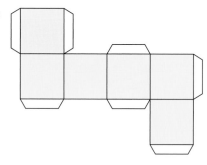

b)

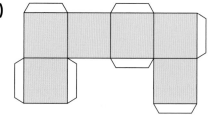

c)

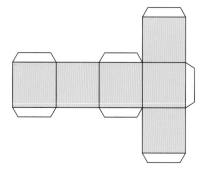

EXERCISE 10.2

In this exercise, look at each net and say what shape it makes.
Then make up the net to check your answer.

1

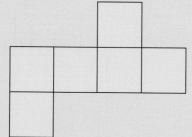

4

2

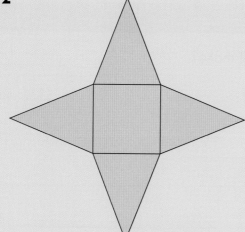

5

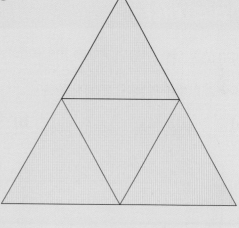

3

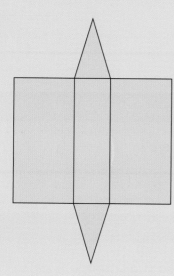

6

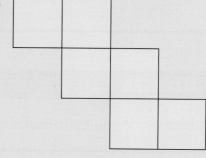

EXERCISE 10.2 continued

7

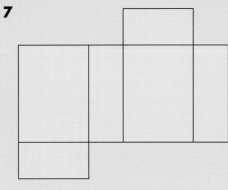

8

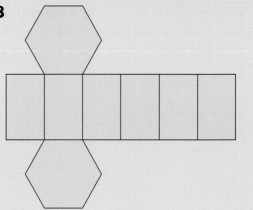

CHALLENGE 1

Can you see what shapes these nets will make?

You can make them if you wish.

a)

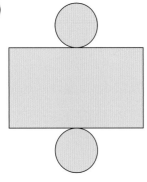

b)

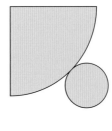

CHALLENGE 2

What happens if you try to make a net for a sphere?

KEY IDEAS

■ A net is a flat shape that can be folded to make a 3-D shape.

Mode and median

You will learn about

- The averages 'mode' and 'median'

You should already know

- How to order numbers

The mode and the median are two different types of 'average'.

The **mode** is the most common number in a set of numbers.

A set of data may have more than one mode or it may have no mode.

The **median** is the middle number of a set of numbers arranged in order.

STAGE
2

245

EXAMPLE 1

Jane and Winston kept a record of their scores on the weekly mental maths test. These are their scores.

Jane	7	8	5	4	7
Winston	10	10	2	1	6

Find the mode and the median for each of them.

The mode is the most common score.

Jane's mode is 7.
Winston's mode is 10.

The median is the middle score when the scores are put in order.

Jane	4	5	7	7	8
Winston	1	2	6	10	10

The shaded scores are the middle scores.

Jane's median score is 7. Winston's median score is 6.

EXAMPLE 2

Here is a list of the weights of people in a 'Keep Fit' class.

73 kg, 58 kg, 61 kg, 43 kg, 81 kg, 53 kg, 73 kg, 70 kg, 73 kg, 62 kg, 60 kg, 85 kg

The instructor wants to know the average weight for the group.

One way is to find the most common weight, the mode.

The mode is 73 kg.

Another way is to put the weights in order and find the middle weight, that is the median.

43 kg, 53 kg, 58 kg, 60 kg, 61 kg, 62 kg, 70 kg, 73 kg, 73 kg, 73 kg, 81 kg, 85 kg

There are two middle weights, 62 kg and 70 kg. The median is halfway between them, 66 kg.

ACTIVITY 1

a) Copy and complete the table below for each of these sets of data.

Data set A   1   1   2   3   3   3   4   4   5   6   7
Data set B   1   1   2   2   3   3   3   4   5   6   7
Data set C   2   2   4   4   6   6   6   8   10   12   14

	Data set A	Data set B	Data set C
Mode			
Median			

b) Write down anything that you notice.

ACTIVITY 2

a) Find the mode and the median for each of these sets of data.
(i)   1   2   3   3   4   5
(ii)   10   20   20   30   70
(iii) 110   120   120   130   170
(iv) 7   10   13   16   19

b) What do you notice about your answers to **(ii)** and **(iii)**?

EXERCISE 11.1

1 Find the median mark for each of these tests.
Remember to write the marks in order, smallest to largest, first.
a) 3   5   6   7   9
b) 8   9   4   3   7   3   1   7
c) 7   8   8   8   8   7   7   8   8   8   7

2 The time taken for a bus journey depends on the time of day.
Here are the times.
15 minutes   7 minutes   9 minutes
12 minutes   9 minutes   19 minutes
6 minutes   11 minutes   9 minutes
14 minutes
a) What is the mode of the times for the journey?
b) What is the median time?

EXERCISE 11.1 continued

3 Twelve people have their handspan measured.
These are the results.

225 mm	216 mm	188 mm
212 mm	205 mm	198 mm
194 mm	180 mm	194 mm
198 mm	200 mm	194 mm

a) How many of the group had a handspan greater than 200 mm?

b) What are the mode and the median?

4 In a survey, a group of boys and girls wrote down how many hours of TV they watch each week.

Boys	Girls
17	9
22	13
21	15
23	17
16	10
12	12
15	12
0	9
5	8
13	12
15	14
13	15
14	
20	

a) Find the mode and the median for each set of figures.

b) Do the boys watch more TV than the girls?

5 Julian counted the matches in ten different matchboxes.
These are the numbers.

48 50 47 50 46
50 49 49 47 50

a) Find the mode.

b) Find the median.

6 These are the earnings of ten workers in a small company.

£10 000	£10 000	£10 000
£10 000	£13 000	£13 000
£15 000	£21 000	£23 000
£70 000		

a) Find the mode and the median for the data.

b) Which of these averages do you think gives the best impression of the average pay?
Explain your answer.

7 These are the ages of a group of people.

19 23 53 19 19
16 26 77 19 27

Find the mode and the median.

8 These are the marks scored in a test.

20 16 18 17 16 18 14 13
18 18 15 18 19 9 12 13

a) Find the mode.

b) Find the median.

EXERCISE 11.1 continued

9 A gardener measures the heights of a group of plants.
These are the heights.

50 cm 65 cm 80 cm 40 cm 35 cm

Find the mode and the median.

10 Seven people go to an evening class to learn how to paint with oils.
These are their ages.

18 19 17 45 37 69 23

Calculate the mode and the median.

CHALLENGE 1

Here are the recent scores of two basketball players.

| Harvey | 24 | 0 | 0 | 32 | 0 | 17 | 29 | 19 | 23 | |
| Nick | 12 | 9 | 3 | 16 | 8 | 6 | 9 | 0 | 11 | 13 |

Work out the mode and the median for each player.

Who would you pick for the next match? Explain your answer.

KEY IDEAS

■ The mode is the most common number in a set of numbers.

■ The median is the middle number of a set of numbers arranged in order.
If there are two middle numbers, the median is halfway between them.

STAGE
2

Formulae

You should already know

- How to add, subtract, multiply and divide whole numbers and decimals
- How to add and subtract negative numbers

| EXAMPLE 1

To find the number of rails needed for a fence, subtract 1 from the number of posts then multiply by 3.

Work out how many rails are needed for a fence with
a) 6 posts. **b)** 25 posts.

a) 6 − 1 = 5
  5 × 3 = 15 rails

b) 25 − 1 = 24
  24 × 3 = 72 rails

EXAMPLE 2

To work out the charge in pounds (£) to hire a concrete mixer, multiply the number of hours by 8 and add on 20.

How much does it cost to hire a concrete mixer for
a) 5 hours? **b)** 25 hours?

a) 5 × 8 + 20 = 40 + 20 = £60

b) 25 × 8 + 20 = 200 + 20 = £220

EXERCISE 12.1

1 This is a rough rule for changing inches into centimetres.

To change inches into centimetres multiply by $2\frac{1}{2}$.

a) About how many centimetres is 2 inches?
b) A foot is 12 inches. About how many centimetres is this?

2 A cookery book gives this rule for cooking chicken.

Allow 15 minutes for each pound (lb) plus another 30 minutes.

a) How long will a 3 lb chicken take to cook?
b) Jill bought a 4 lb chicken. She put it in the oven at 4 p.m.
When will the chicken be ready?

3 Mach number is a measure of speed.
An aeroplane travelling at Mach 1 is travelling at the speed of sound.
There is a rule for working out the Mach number.

Mach number = speed in miles per hour ÷ 760

Calculate the Mach number of a plane travelling at 1520 mph.

4 There is a rule for finding how far away a thunderstorm is.

Count the seconds from the lightning to the thunder. Divide by 5.
The answer is the distance in miles.

a) How far away is the storm if you count 5 seconds?
b) How far away would it be if you counted 30 seconds?

STAGE
2

EXERCISE 12.1 continued

5 To work out her weekly pay Jodi uses this formula.

> Weekly pay = rate of pay per hour × hours worked + bonus

One week Jodi works for 30 hours at a rate of £6 per hour and earns a bonus of £30.
Calculate her pay.

6 As you go further up a mountain, it gets colder.
There is a simple formula which tells you roughly how much the temperature will drop.

> Temperature drop (°C) = height climbed in metres ÷ 200

If you climb up 800 m, about how much will the temperature drop?

7 Here is a rough rule to convert gallons into litres.

> Multiply the number of gallons by 9 and divide by 2.

a) About how many litres is 200 gallons?
b) About how many litres is 350 gallons?

8 To work out the number of rolls of wallpaper needed to paper a room some DIY books give this rule.

> Measure the distance round the edge of the room in feet. Call this number D.
> Measure the height of the room in feet. Call this number H.
> Multiply these two numbers together, i.e. find $D \times H$.
> Now divide by 50.
> The answer gives the number of rolls needed.

a) The distance round the sitting room is 65 feet and the height is 10 feet. How many rolls are needed?
b) The distance round a bedroom is 50 feet and the height is 8 feet. How many rolls are needed?

9 Here is a rule to find the volume of a cone.

> Multiply the area of the base by the height and then divide by 3.

Find the volume of a cone when the area of the base is 20 cm² and the height is 12 cm.

10 This rule gives the approximate distance round a circle.

> Multiply the radius by 6.

a) A circle's radius is 6 cm. What is the distance round it?
b) A round cake has a radius of 8 cm.
   What length of ribbon is needed to just go round it?

11 Here is a rule to change from pounds (£) to euros.

> Multiply the number of pounds by 3 and then divide by 2.

How many euros would you get for £120?

12 Here is a rule to change from miles per hour to metres per second.

> Multiply by 4 and divide by 9.

a) Change 45 miles per hour into metres per second.
b) Change 27 miles per hour into metres per second.

C CHALLENGE 1

Here are two formulae for changing a temperature in Celsius (°C) into Fahrenheit (°F).

■ **Approximate formula**
Multiply the temperature in Celsius by 2 and then add 30.

■ **Accurate formula**
Multiply the temperature in Celsius by 9, divide by 5 and then add 32.

a) Use each formula to find the Fahrenheit temperature equal to 20°C.
b) Find the temperature which gives the same result with both formulae.

STAGE 2

C CHALLENGE 2

Here are some word formulae for you to use. (*Formulae* is the plural of *formula*.)

Length in millimetres is the length in centimetres multiplied by 10.

The amount in kilograms is roughly the amount in pounds divided by 2.

Distance travelled is equal to the speed multiplied by the time taken.

Use these formulae to change or calculate each of these.

a) 30 cm to millimetres

b) 10 pounds to kilograms

c) The distance travelled by a car going at 40 mph for 4 hours

d) 60 mm to centimetres

e) 40 kilograms to pounds

f) The time taken to travel 18 miles at 3 miles per hour

K KEY IDEAS

■ A word formula shows how to work out a problem.

Revision exercise C1

1 This angle measures 50°.

50°

Estimate the size of each of these angles.

Write down your estimates and then check them using a protractor or angle measurer.

a)

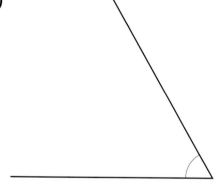

b)

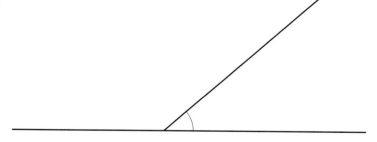

c)

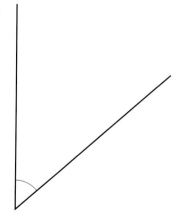

2 The leopard is 1·5 m tall.

Estimate the height of the elephant and the giraffe.

3 This line is 6 cm long.

_____

Use this line to estimate the length of each of these lines.

a) ─────────

b) ─────────────────────────────

c) ───────────────────────────────────────

4 Here are four 3-D shapes and their nets.
Match each 3-D shape to its net and name the 3-D shape.

a) **b)** **c)** **d)**

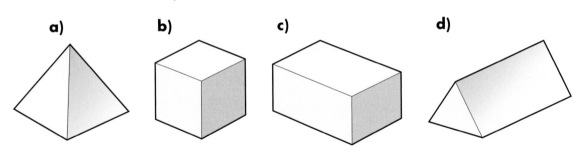

(i) **(ii)** **(iii)** **(iv)**

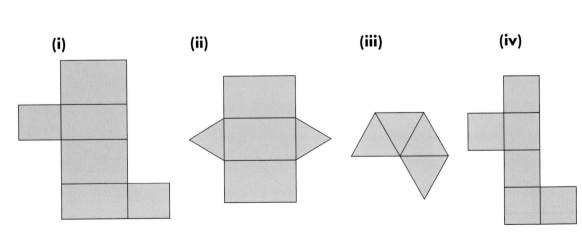

5 Find the mode and the median of each of these sets of data.
  a) 4   3   15   9   7   6   11   6   12
  b) 60 kg   12 kg   48 kg   36 kg   24 kg

6 A gardener measures the height of his sunflower plants.
These are the heights.

140 cm   123 cm   131 cm   89 cm   125 cm   123 cm   115 cm   138 cm

Find the mode and the median of these heights.

7 The cost of hiring a carpet cleaner is £15 deposit plus £4 per day.
How much does it cost to hire the cleaner for four days?

8 A travel agent uses this formula to change pounds (£) into dollars ($).

Multiply the number of pounds by 15 and divide by 10.

Calculate how many dollars you would get for these amounts.
  a) £10
  b) £40

13 Reflection symmetry

Recognising reflection symmetry

Emma is moving into a new bungalow.

The removal van is covering exactly half of the front of the bungalow.
The hidden part is the same as the part you can see, only in reverse.

What will the whole front of the bungalow look like?

You can place a small mirror along the line of the back of the van to find out.

You will see the other half of the bungalow in the reflection.

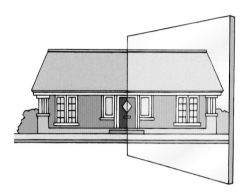

As the two halves are mirror images, we say the bungalow has **reflection symmetry**.

The line that divides the two parts is called the **line of symmetry** or the **mirror line**.

A shape may have more than one line of symmetry.

EXAMPLE 1

Which of these shapes have reflection symmetry?

For those that have, draw in the line(s) of symmetry.

If you put a mirror down the middle of each shape, you can see which have symmetry.

You may need to move the mirror to find the position of the line(s).

But, however hard you try, you cannot find lines for one of the shapes.

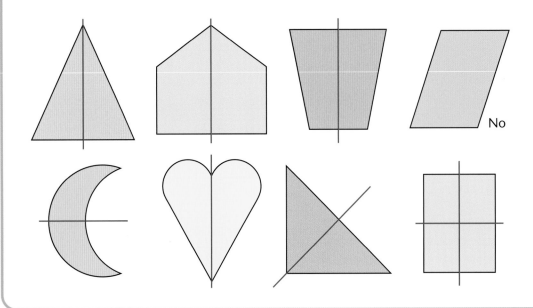

EXERCISE 13.1

Which of these shapes have reflection symmetry?
For those that have, copy the shape and draw in the line(s) of symmetry.

1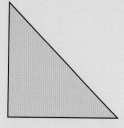

2

3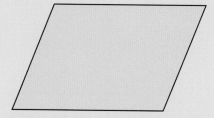

4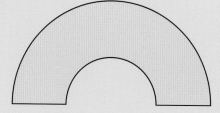

5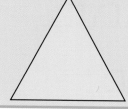

6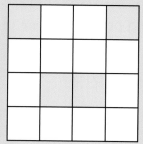

7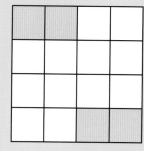

8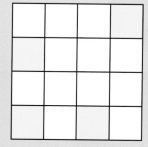

9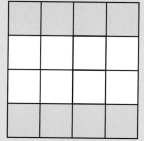

STAGE
2

Drawing reflections

Look again at the picture of the bungalow on page 258.

You can draw the reflection using tracing paper.

Place the tracing paper with one of its edges along the line of the back of the van.

Trace round the bungalow.

Turn the tracing paper over, keeping the edges lined up, and this will complete the front of the bungalow.

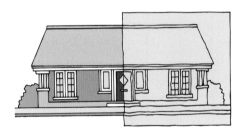

Reflection writing

EXERCISE 13.2

Copy each of these and complete it so it has reflection symmetry in the red mirror line shown.
Each one spells a word, gives a number or is a calculation.

1 DED

2 HIDE

3 ICE

4 KICK

5 CODE

6 HOD

EXERCISE 13.2 continued

7 KID

8 CHOKE

9 BOX

10 HOOD

11 10

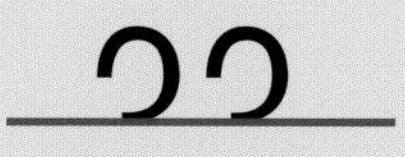

12 22

13 001

14 oo

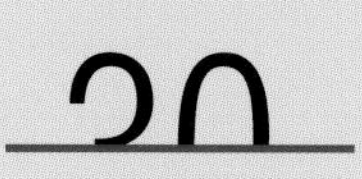

15 101

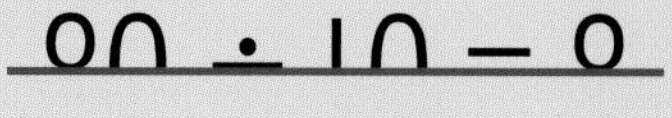

16 30

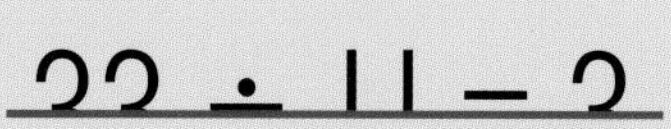

17 10 + 12 = 21

18 00 ÷ 10 = 0

19 22 ÷ 11 = 2

20 0 × 0 = 0

A ACTIVITY 1

See if you can find some more reflection words, numbers or calculations.

STAGE
2

Simple reflections

This section involves completing patterns on squared paper so that they have reflection symmetry.

EXAMPLE 2

Complete the shape so that it has reflection symmetry in the mirror line shown.

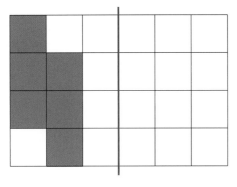

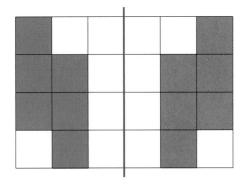

EXAM TIP

You can use tracing paper to help. In class you could check your answer with a small mirror.

There are two things to notice.

■ The shape is 'turned over' in a reflection.

■ The 'image' is the same distance from the mirror line as the original shape but is on the opposite side of the line.

EXERCISE 13.3

Copy these shapes on to squared paper.
Complete each of the shapes so that it has reflection symmetry in the mirror line shown.

1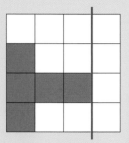

2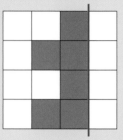

3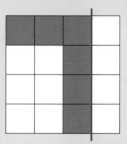

4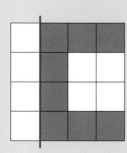

5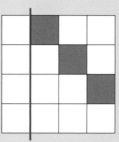

6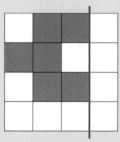

7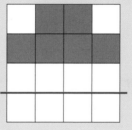

8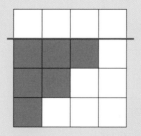

STAGE
2

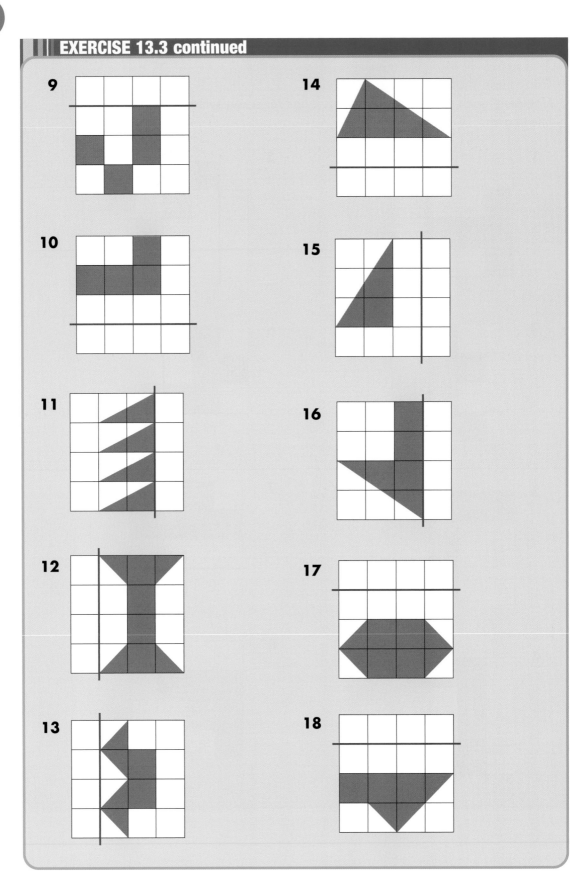

9

10

11

12

13

14

15

16

17

18

19

20

More difficult reflections

Type 1

Sometimes the original shape may have separate parts on both sides of the mirror line.

In these cases, reflect each part separately on to the other side of the mirror line.

EXAMPLE 3

Complete this shape so that it has reflection symmetry in the mirror line shown.

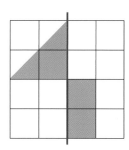

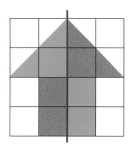

Type 2

Sometimes you may be asked to complete a shape so that it has reflection symmetry in two mirror lines which cross at right angles.

■ Reflect the given shape in one of the mirror lines.

■ Then reflect the result in the second mirror line.

EXAMPLE 4

Complete this simple shape so that is has reflection symmetry in both mirror lines shown.

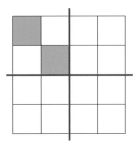

Reflect the given shape in one of the mirror lines.

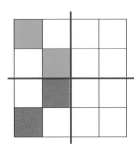

Then reflect the result in the second mirror line.

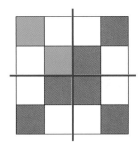

EXERCISE 13.4

Copy these shapes on to squared paper.
Complete each of the shapes so that it has reflection symmetry in the mirror line(s) shown.

1

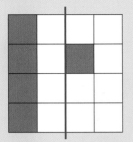

2

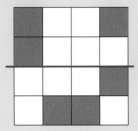

3

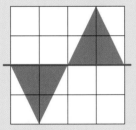

4

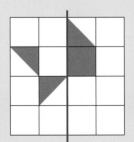

5

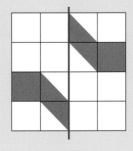

6

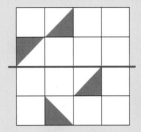

7

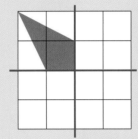

8

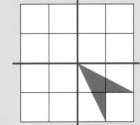

9

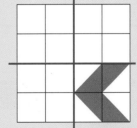

10

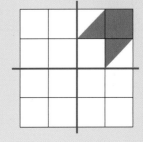

STAGE
2

C CHALLENGE 1

Copy these shapes on to squared paper.

Complete each of the shapes so that it has reflection symmetry in the mirror line(s) shown.

a)
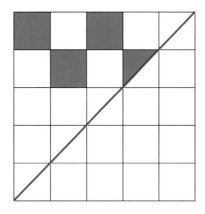

b)
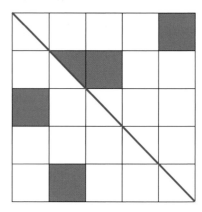

c)
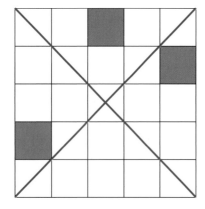

K KEY IDEAS

■ 2-D shapes may be reflected in a mirror line. The mirror line is a line of symmetry for the complete new shape.

Maps and plans

You will learn about

- Grid references on maps
- Giving and following directions on a map

You should already know

- How to read scales and coordinates
- About compass directions

Grid references

Maps and plans are often divided into grids. Each grid square is identified with a number to help people to find the position of objects. These numbers are called the **grid references** for the squares and are given by the left-hand bottom corner, the east number first and then the north number.

Maps usually show the direction of north, and sometimes also the directions of south, east and west.

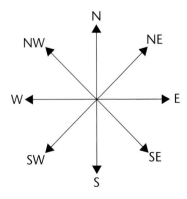

EXAMPLE 1

Here is a map showing the position of some ships near a port.

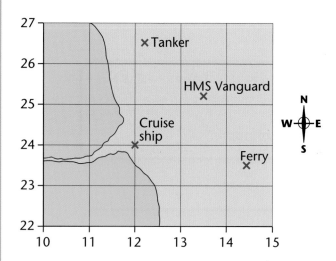

a) Write down the grid reference for each of the ships.

b) From HMS *Vanguard* in which direction is
   (i) the ferry?
   (ii) the tanker?

c) A coastguard boat goes from the cruise ship to HMS *Vanguard* and on to the tanker.
   Does it travel clockwise or anticlockwise?

a) The cruise ship is on the left-hand bottom corner of a square.
   Its grid reference is 12 24.

   Find the square with the tanker in.
   The left-hand bottom corner of the square is at 12 26, so this is the grid reference of the tanker.

   The grid reference of *HMS Vanguard* is 13 25 and of the ferry is 14 23.

b) From *HMS Vanguard*
   (i) the ferry is south east.
   (ii) the tanker is north west.

c) Anticlockwise

EXERCISE 14.1

1 Here is part of a map. It shows the position of two roads and some nearby landmarks.

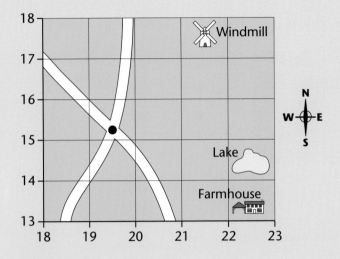

a) Write down the grid reference of each of these.
   (i) The roundabout
   (ii) The windmill
   (iii) The farmhouse
b) The farm is due south of the lake. In which direction is the lake from the farm?
c) Write down the direction of the roundabout from the windmill.

2 This map shows a park.
   a) What is at 15 41?
   b) Write down the grid reference of each of these.
   (i) The tennis court
   (ii) The seat
   c) Copy and complete this sentence with compass directions.
   The fountain is
   .... of the pond
   and .... of the
   greenhouse.

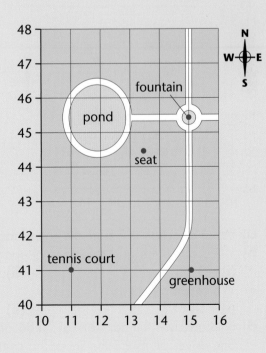

3 This map shows an island.

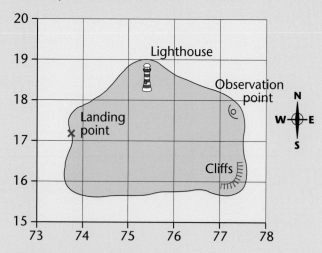

a) Write down the grid reference of each of these.
   (i) The landing point
   (ii) The lighthouse
   (iii) The observation point
b) Harry starts from the landing point and sails round the island in a clockwise direction. Name the places he passes, in order.

4 This is the plan of a shopping centre.

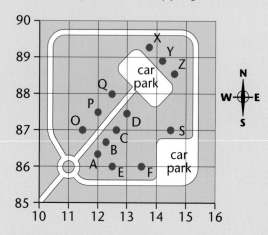

a) What direction is F from the roundabout?
b) Write down the grid reference of each of these.
   (i) S   **(ii)** P
c) Which shop is at
   (i) 11 87?   **(ii)** 13 89?
d) Sheila leaves the car park at 15 86 and drives north.
   She drives all round the shopping centre.
   Which direction is this, clockwise or anticlockwise?

5 This map shows part of a park.

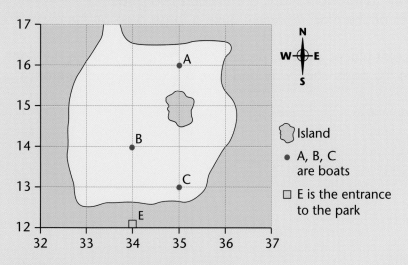

a) What is due south of the island?

b) What is at 34 12?

c) Give the grid references of boats B and C.

6 The map shows part of Northern Ireland.

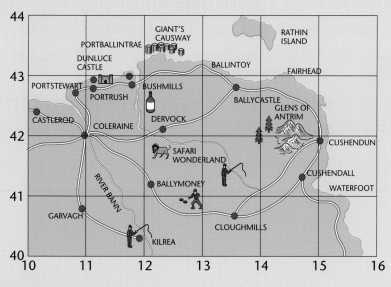

a) Write down the grid reference of each of these.

   (i) Portrush

   (ii) Cushendall

   (iii) Cloughmills

b) **(i)** Which town is almost due west of Dervock?

   (ii) Which tourist attraction is north east of Garvagh?

Ordnance Survey *Landranger* maps are divided into grids.
The grid squares on the map represent quite a large area on the ground –
1 km each way.
So to help pinpoint places exactly, people often imagine that the divisions
are divided into 10 parts each way and give a six-figure grid reference.

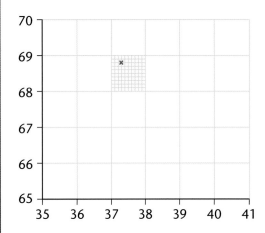

Look at the cross marked on the grid. It is about $\frac{3}{10}$ across from the 37 grid
line and about $\frac{8}{10}$ up from the 68 grid line. The six-figure grid reference is
373 688. Remember that the first three figures refer to the east direction
and the second three to the north direction.

Working with a partner, find a place on an Ordnance Survey map and
write down its grid reference.

Ask your partner to find the point. Now change over.

Using maps

Are you lost? You need a map!

A ACTIVITY 1

The map shows part of London.

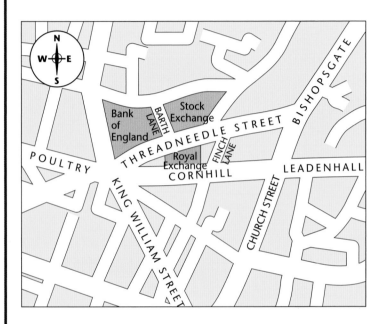

Find the Bank of England and follow these directions.

- Leave the bank and turn LEFT along Threadneedle Street.

- Pass a street on your left. What is it called?

- What is the next building on your left?

- Take the next turn RIGHT. What street are you in?

- Take the next RIGHT. In which direction are you walking?

- What building is on your right?

- Return to the Bank of England.
  In which direction around the streets was this route?

EXERCISE 14.2

1 This map shows part of Stratford-upon-Avon. Copy and complete these directions.

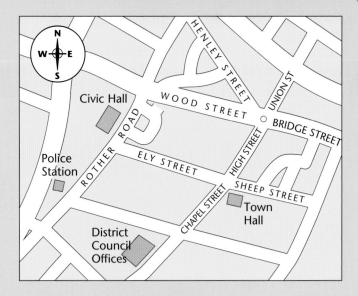

Leave the Civic Hall and turn LEFT along Rother Road.
At the crossroads turn RIGHT along ....
Take the next RIGHT into ....
You are walking .... (compass direction).
At the next crossroads .... is on the corner.

2 This map shows part of Plymouth.
Copy and complete these directions.

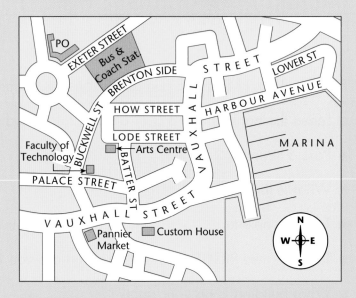

Turn LEFT out of the Bus Station on Brenton Side.
Turn .... into Vauxhall Street.
Take the next LEFT into ....
Go straight on, passing the .... on your RIGHT.

3 Here is a map of part of Harrogate.

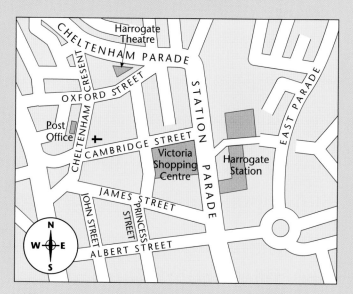

Copy and complete these directions.
Turn RIGHT out of Harrogate Station and walk north along ....
Turn .... into Cambridge Street.
Take the second on the RIGHT into .... and go into
.... on your LEFT.

4 Here is a map of part of Stoke-on-Trent. Copy and complete these directions.

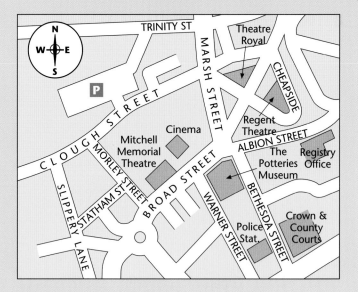

Leave the Car Park on Clough Street and turn RIGHT.
Take the first LEFT into ....
At the end of the street, turn LEFT into ....
You are now walking.................... (compass direction).
Take the third on the RIGHT into .... and pass
.... on your LEFT.

5 This is a map showing part of St Andrews. Write out a route from the Cinema to the Town Hall.

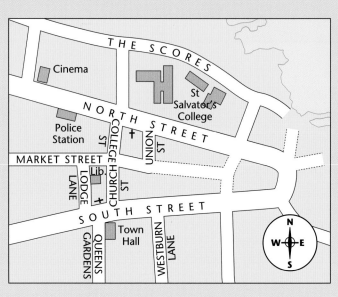

EXERCISE 14.2 continued

6 This is a map of the centre of Huddersfield.

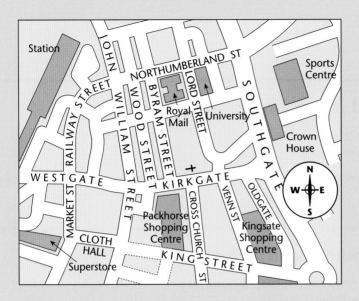

Write a route from the Station to the Kingsgate Shopping Centre.

KEY IDEAS

■ Grid references are given with the east number first.

■ On maps, directions can be given using compass directions, or as turning left or right.

15 Two-way tables

Using two-way tables

Two-way tables are very useful for displaying complicated information.

EXAMPLE 1

The girls in Year 11 have chosen whether they want to play hockey and/or netball during their PE lessons.

The information is shown in this two-way table.

	Hockey	Not hockey	Total
Netball	35	20	55
Not netball	15	10	25
Total	50	30	80

a) How many girls are there altogether?

b) How many want to play hockey?

c) How many want to play hockey but not netball?

d) How many do not want to play either game?

a) 80 This is at the bottom right-hand corner.

b) 50 This is the total for the Hockey column (35 + 15).

c) 15 You find this by looking in the Hockey column and the Not netball row.

d) 10 This is the entry for Not hockey and Not netball.

STAGE
2

EXERCISE 15.1

1 Here is a two-way table showing the results of a survey looking at the makes and colours of cars.

	Vauxhall	Not Vauxhall	Total
Black	20	50	70
Not black	60	300	360
Total	80	350	430

a) How many cars are black but are not Vauxhalls?
b) How many cars are black?
c) How many cars are Vauxhalls?

2 A drugs company has compared a new type of drug for hay fever with an existing drug.
Here is a two-way table showing the results of the trial.

	Existing drug	New drug	Total
Symptoms eased	700	550	1250
No change in symptoms	350	250	600
Total	1050	800	1850

a) How many people took part in the trial?
b) How many people using the new drug had their symptoms eased?
c) How many people had their symptoms eased?

3 At the world indoor athletics championships the USA, Germany and China won most medals.
This is the medal table for these three countries.

	Gold	Silver	Bronze	Total
USA	31	18	10	59
Germany	18	16	9	43
China	22	9	11	42
Total	71	43	30	

a) One box has not been completed.
　　(i) What should the number be?　　**(ii)** What does it represent?
b) Which of these countries won the most gold medals?
c) Which of these countries won the most bronze medals?

EXERCISE 15.1 continued

4 This distance chart shows the distances in miles between various towns in Wales.

a) How far is it from Newport to Swansea?

b) How far is it from Cardiff to Neath?

c) Morgan drove from Cardiff to Swansea to Carmarthen and directly back to Cardiff. How far did he drive altogether?

Distance (miles)

Cardiff				
64	Carmarthen			
36	30	Neath		
91	29	58	Newport	
41	27	10	55	Swansea

5 This distance chart shows the distances in miles between some places in the UK.

Distance (miles)

Norwich							
119	Nottingham						
161	103	Oxford					
423	328	264	Penzance				
405	311	418	611	Perth			
76	58	86	358	351	Peterborough		
354	259	195	78	543	289	Plymouth	
205	197	85	244	502	160	176	Portsmouth

a) How far is it from Perth to Oxford?

b) Which two places in the table are furthest apart?

c) Kate drove from Nottingham to Peterborough and then on to Portsmouth. How far did she drive altogether?

STAGE
2

6 The table shows the car hire rates of a company in Portugal.

Group	Price per week in		
	Jan/Feb/ March/Nov/Dec	April/May/ June/Oct	July/Aug/ Sept
A Renault Clio/ Citroen C3 3 doors	€130	€150	€190
B Renault Clio/ Citroen C3 5 doors	€140	€160	€200
C Renault Clio/ Citroen C3 5 doors with Air Con.	€160	€200	€250
D Megane/Xsara/ Shuma/Lanos or similar. Air Con.	€235	€275	€320
E Citroen Xsara/ Renault Megan estate or similar	€320	€285	€340
F Bus/ VW Transporter	€310	€410	€510
G Monovolume 7 seater	€410	€510	€610

a) Harry hires a Renault Clio with three doors for a week in April.
   How much does it cost?
b) The Brown family hire a Citroen C3 with five doors and air conditioning
   for two weeks in August.
   How much does it cost?
c) How much more does it cost to hire a Renault Megan Estate for two weeks
   in July than for two weeks in June?

C CHALLENGE 1

A group of boys were asked whether or not they play cricket or football.

Some of the results are shown in this table.

	Cricket	Not cricket	Total
Football		18	
Not football	15		
Total		25	48

Copy and complete the table.

Timetables

What time will a bus come?

It sometimes feels as if they come whenever they feel like it, especially when two come together!

However, they are trying to run to a **timetable**.

15

Look at this part of a simplified train timetable.

Each **row** represents one train.

London	Long Eaton	Derby
0610	0755	0812
0800	0943	0956
1200	1343	1356
1715	1855	1909

There are two things to notice.

■ The times all have four digits, for example 0943 means 9.43 a.m.

■ After 12 o'clock (1200) the times do not go back to 1 again.
For example, 1.43 p.m has become 1343, by adding 12 to the hour.

These times are in the **24-hour clock**.

There are 24 hours in the day so this means we do not need a.m. (for morning) and p.m. (for afternoon and evening).

> **EXAM TIP**
> 24-hour times are sometimes written with a colon ':', for example 13:43.

EXAMPLE 2

Look at the train timetable above.
I am in London and I want to get to Derby before 10 a.m.

a) Which is the best train to catch?

b) How long will the journey take?

a) To arrive at Derby before 10 a.m., that is 1000, the best train to catch is the 0800 train, which arrives at 0956, just in time.

b) It is 1 hour from 0800 to 0900 and another 56 minutes until the train arrives.
The total journey time is 1 hour 56 minutes.

> **EXAM TIP**
> Remember that there are 60 minutes in an hour.
> Do not use your calculator as you may well get the wrong answer.

Some timetables are written the other way.

In the timetable in Example 3, for example, each **column** represents one bus.

EXAMPLE 3

Here is part of a bus timetable.

Derby Bus Station	0545	0600	0615	0630
City General Hospital	0555	0610	0625	0640
Square	0601	0616	0631	0646
Ladybank Road	0606	0621	0636	0651
East Avenue	0610	0625	0640	0655

a) How long does it take to travel from Derby Bus Station to Ladybank Road?

b) When will I reach East Avenue if I catch the 0625 bus from the City General Hospital?

a) Each bus takes the same time.
0545 to 0606 is 15 + 6 minutes = 21 minutes.
You should check that the other buses are the same.
Which is the easiest to calculate?

b) 0640

EXERCISE 15.2

1 Look at the bus timetable in Example 3.
The pattern of times continues for the rest of the day.
Copy and complete the timetable for the next bus.

Derby Bus Station	
City General Hospital	
Square	
Ladybank Road	
East Avenue	

EXERCISE 15.2 continued

2 Trains from London to Derby go through Leicester.
They take 1 hour 25 minutes to reach Leicester.
Copy and complete the timetable to show when each train arrives at Leicester.

London	Leicester	Derby
0610		0812
0800		0956
1200		1356
1715		1909

3 This is part of a Swiss railway timetable.

Brig	Spiez	Bern	Solothurn
0618	0722	0754	0843
0652	0748	0821	0913
0818	0922	0954	1043
1052	1148	1221	1313
1518	1622	1654	1743
1718	1822	1854	1943
1903	1949	2021	2154
2118	2222	2254	2354

a) If you catch the 0922 train from Spiez, when will you get to Solothurn?
b) How long does that journey take?
c) Which is the best train to catch in Brig if you want to reach Bern by
7 p.m.?
d) Herr Schmitt arrived at Solothurn at 1743.
Which train did he catch from Spiez?
e) How long was his journey?
f) How long does the 1052 train from Brig take to reach Solothurn?

EXERCISE 15.2 continued

4 Here is a timetable for flights between Aberdeen and London.

Flight	Depart	Arrive
BA 101	0630	0810
BA 105	0940	1115
BA 107	1255	1425
BA 109	1405	1535
BA 113	1715	1840
BA 115	1835	2010
BA 117	2010	2145

a) Which is the best flight to catch if you want to reach London before 2 p.m.?
b) How long is flight BA 109?
c) You are ready to leave Aberdeen at 1 p.m.
What is the earliest time you can arrive in London?
d) Do all the flights take the same time?
If not, work out the different times.

5 Here is the timetable for some trains running on the line between Leeds and Sheffield.

Leeds	0828	0916	1016	1316	1716
Wakefield	0839			1323	
York		0927	1034	1328	1733
Barnsley		0948	1048	1348	1748
Doncaster		0956	1058	1353	1757
Meadowhall		1007	1107	1403	1807
Sheffield	0927	1016	1116	1412	1816

a) Find the train which arrives at Sheffield at 9.27 a.m.
Explain why the timetable does not give a time for York, Barnsley, Doncaster or Meadowhall.
b) What can you say about some trains and Wakefield?
c) How long does the 0927 train from York take to reach Sheffield?
d) Which train is the quickest between Leeds and Sheffield?
How long does it take?

STAGE
2

C   CHALLENGE 2

Using local timetables for buses and trains, plan a day trip from home. Make a timetable for the trip.

Don't forget to allow enough time to make the changes, and don't miss the last train home!

K   KEY IDEAS

- Two-way tables can be used to display complicated information.

- Timetables are two-way tables showing the times of events.

Revision exercise D1

1 Which of these shapes have reflection symmetry?
For those that have, copy the shape and draw in the line(s) of symmetry.

a)

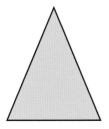

b)

c)

d)

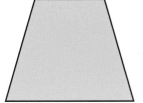

e)

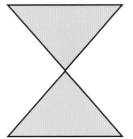

f)

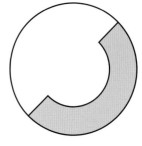

2 Copy this shape on to squared paper.
Complete the shape so that it has reflection symmetry in the mirror line shown.

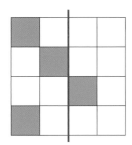

3 Copy these shapes on to squared paper.
Complete each of the shapes so that it has reflection symmetry in the mirror lines shown.

a)

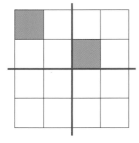

b)

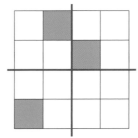

c)

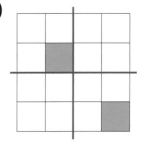

4 This map shows part of a coast where a river enters the sea.

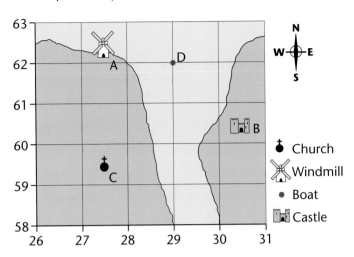

a) What is due west of the boat at D?
b) What is the grid reference of the castle at B?
c) What is at 27 59?
d) Write down the grid reference of each of these.
   (i) The windmill at A
   (ii) The boat at D

5 This is a map of part of Sheffield.

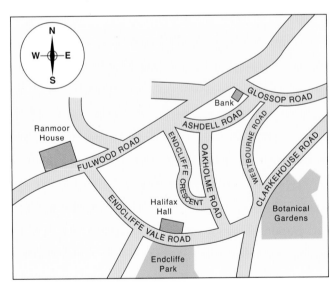

a) I leave Ranmoor House and turn left.
   I take the first road on the right.
   What do I pass on the left down this road?
b) Write directions to go from the bank on Glossop Road to the Botanical Gardens.

6 Year 7 is going to have a day out at the end of the term.
The students were asked to vote for their chosen place.
The table shows their votes.

	Alton Towers	Legoland	Thorpe Park	Total
Girls	31	25	27	83
Boys	43	11	21	75
Total	74	36	48	158

a) How many boys wanted to go to Legoland?
b) How many girls wanted to go to Thorpe Park?
c) How many students voted?
d) The day out was at the place that received the most votes.
Where was the day out?

7 This timetable is for buses between Burton-on-Trent and Derby.

Burton-on-Trent Station	1708	1808	1857	1957	2157
High Street	1711	1811	1904	2004	2204
Repton	1726	1826	1917	2017	2217
Willington	1735	1830	1920	2020	2220
Findern	1740	1835	1925	2025	2225
Mickleover	1747	1842	1930	2030	2230
Littleover	1751	1846	1934	2034	2234
Derby	1805	1900	1945	2045	2245

a) Mrs Jones got off the bus in Littleover at 1751.
What time did she catch the bus in Repton?
b) How long does it take to go from Burton-on-Trent Station to Derby on the 1708?
c) You need to be in Derby for 2000.
Which bus is the best one to catch from the High Street?

Index